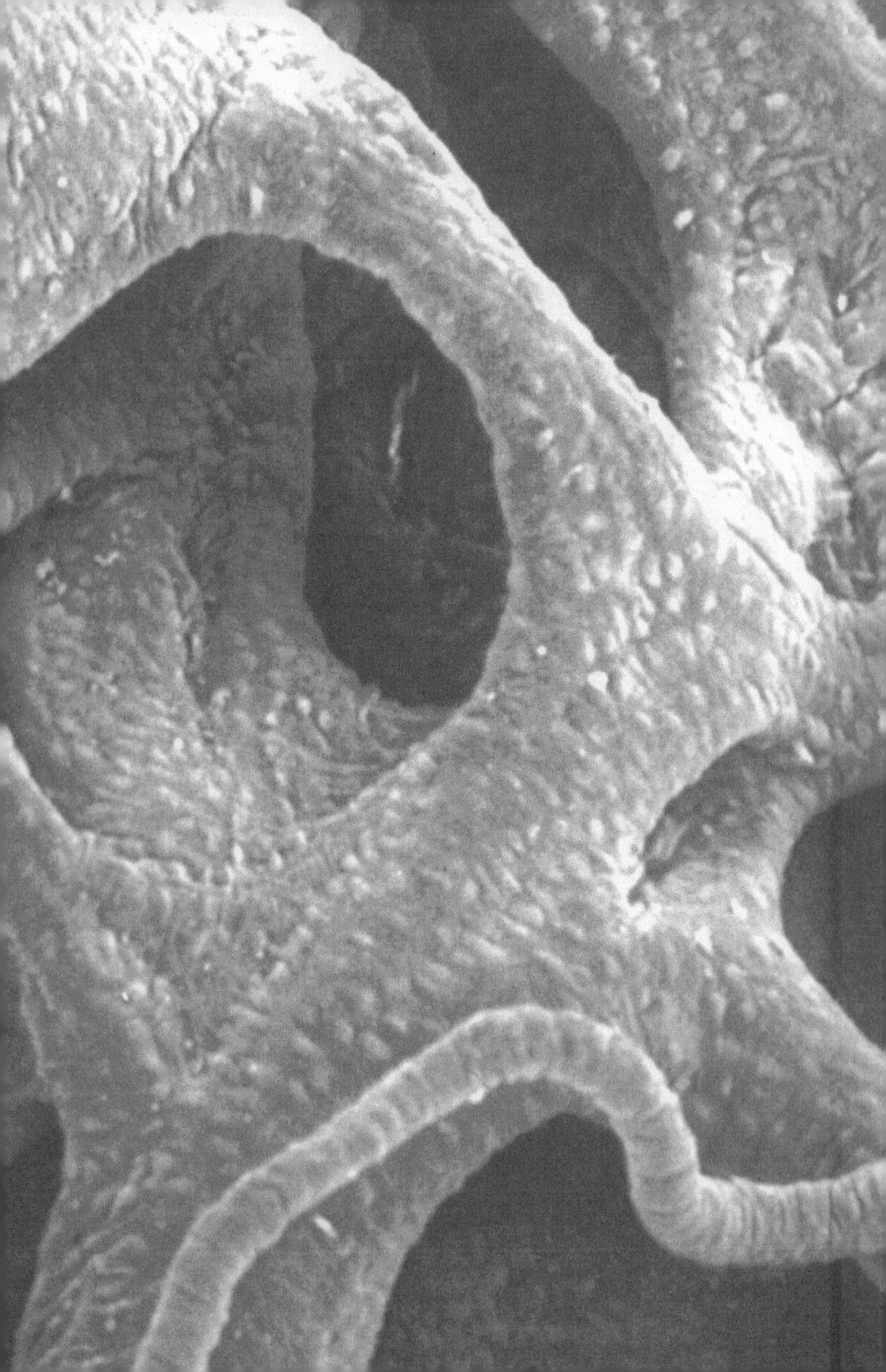

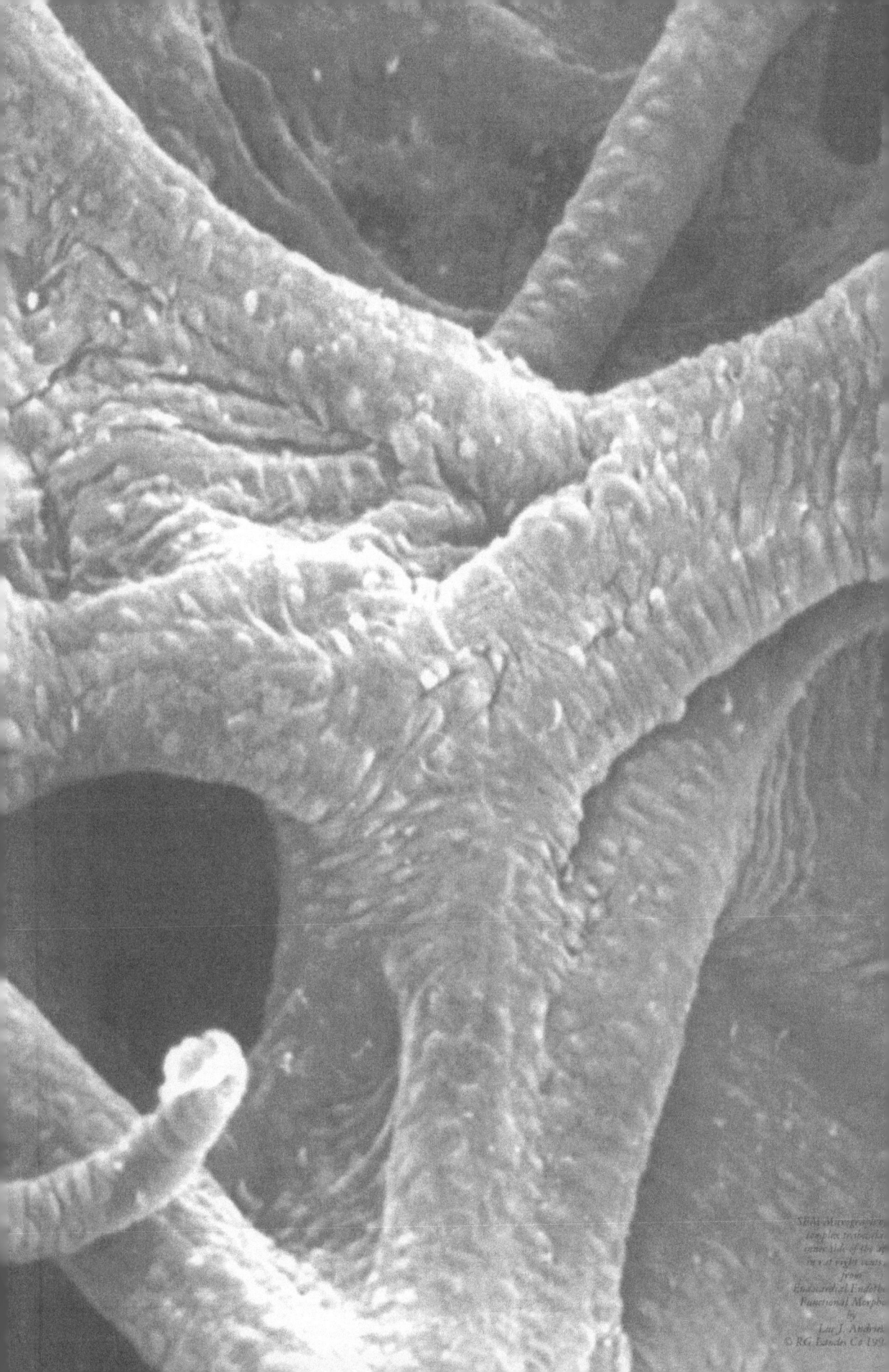

MEDICAL
INTELLIGENCE
UNIT

SIGNAL TRANSDUCTION AND BACTERIAL VIRULENCE

MEDICAL
INTELLIGENCE
UNIT

SIGNAL TRANSDUCTION AND BACTERIAL VIRULENCE

Rino Rappuoli
Vincenzo Scarlato
Beatrice Aricò

Immunobiological Research Institute Siena
Siena, Italy

SPRINGER-VERLAG BERLIN HEIDELBERG GMBH

MEDICAL INTELLIGENCE UNIT

SIGNAL TRANSDUCTION AND BACTERIAL VIRULENCE

International Copyright © 1995 Springer-Verlag Berlin Heidelberg
Originally published by Springer-Verlag in 1995
Softcover reprint of the hardcover 1st edition 1995

International ISBN 978-3-662-22408-3

Library of Congress Cataloging-in-Publication Data

Signal transduction and bacterial virulence / [edited by] Rino Rappuoli.
     p. cm. — (Medical intelligence unit)
   Includes bibliographical references and index.
   ISBN 978-3-662-22408-3   ISBN 978-3-662-22406-9 (eBook)
   DOI 10.1007/978-3-662-22406-9
   1. Virulence (Microbiology) 2. Cellular signal transduction.
   I. Rappuoli, Rino. II. Series.
      [DNLM: 1. Bacteria—pathogenicity. 2. Signal Transduction. QW 730 S578 1995]
QR175.S47 1995   616.'014—dc20
DNLM/DLC   95-6581
for Library of Congress   CIP

Publisher's Note

R.G. Landes Company publishes five book series: *Medical Intelligence Unit, Molecular Biology Intelligence Unit, Neuroscience Intelligence Unit, Tissue Engineering Intelligence Unit* and *Biotechnology Intelligence Unit*. The authors of our books are acknowledged leaders in their fields and the topics are unique. Almost without exception, no other similar books exist on these topics.

Our goal is to publish books in important and rapidly changing areas of medicine for sophisticated researchers and clinicians. To achieve this goal, we have accelerated our publishing program to conform to the fast pace in which information grows in biomedical science. Most of our books are published within 90 to 120 days of receipt of the manuscript. We would like to thank our readers for their continuing interest and welcome any comments or suggestions they may have for future books.

Deborah Muir Molsberry
Publications Director
R.G. Landes Company

CONTENTS

EDITORS

Rino Rappuoli
Immunobiological Research Institute Siena (IRIS)
Siena, Italy
Introduction

Vincenzo Scarlato
Immunobiological Research Institute Siena (IRIS)
Siena, Italy
Introduction

Beatrice Aricò
Immunobiological Research Institute Siena (IRIS)
Siena, Italy
Introduction

CONTRIBUTORS

Brian J. Akerley
Department of Microbiology and
  Immunology
University of California
Los Angeles, California, U.S.A.
Chapter 2

Naomi Balaban
Immunobiological Research
  Institute Siena (IRIS)
Siena, Italy
Introduction

Jutta Bohne
Theodor-Boveri-Institut für
  Biowissenschaften (Biozentrum)
Lehrstuhl für Mikrobiologie
Universität Würzburg
Würzburg, Germany
Chapter 8

Guy R. Cornelis
Microbial Pathogenesis Unit
International Institute of Cellular
  and Molecular Pathology (ICP)
  and Faculté de Médecine
Université Catholique de Louvain
Brussels, Belgium
Chapter 6

Peggy A. Cotter
Department of Microbiology and
  Immunology
University of California
Los Angeles, California, U.S.A.
Chapter 2

V. Deretic
Department of Microbiology
University of Texas Health Science
  Center in San Antonio
San Antonio, Texas, U.S.A.
Chapter 3

CONTRIBUTORS

Victor J. DiRita
Department of Microbiology and
    Immunology and
Unit for Laboratory Animal Medicine
University of Michigan Medical
    School
Ann Arbor, Michigan, U.S.A.
Chapter 5

Werner Goebel
Theodor-Boveri-Institut für
    Biowissenschaften (Biozentrum)
Lehrstuhl für Mikrobiologie
Universität Würzburg
Würzburg, Germany
Chapter 8

Roy Gross
Theodor-Boveri-Institut für
    Biowissenschaften (Biozentrum)
Lehrstuhl für Mikrobiologie
Universität Würzburg
Würzburg, Germany
Chapter 8

Maite Iriarte
Microbial Pathogenesis Unit
International Institute of Cellular and
    Molecular Pathology (ICP) and
    Faculté de Médecine
Université Catholique de Louvain
Brussels, Belgium
Chapter 6

Hubert Kestler
Theodor-Boveri-Institut für
    Biowissenschaften (Biozentrum)
Lehrstuhl für Mikrobiologie
Universität Würzburg
Würzburg, Germany
Chapter 8

Jürgen Kreft
Theodor-Boveri-Institut für
    Biowissenschaften (Biozentrum)
Lehrstuhl für Mikrobiologie
Universität Würzburg
Würzburg, Germany
Chapter 8

Anthony T. Maurelli
Department of Microbiology and
    Immunology
Uniformed Services University
    of the Health Sciences
F. Edward Hebert School of Medicine
Bethesda, Maryland, U.S.A.
Chapter 7

Jeff F. Miller
Department of Microbiology and
    Immunology and
    Molecular Biology Institute
University of California
Los Angeles, California, U.S.A.
Chapter 2

Samuel I. Miller
Infectious Disease Unit
Harvard Medical School
Massachusetts General Hospital
Boston, Massachusetts, U.S.A.
Chapter 4

John R. Murphy
Evans Department of Clinical
    Research and Department
    of Medicine
Boston University Medical Center
    Hospital
Boston, Massachusetts, U.S.A.
Chapter 1

CONTRIBUTORS

Richard P. Novick
Skirball Institute of Biomolecular
    Medicine
New York, New York, U.S.A.
Chapter 9

Catherine M. C. O'Connell
Department of Microbiology and
    Immunology
Uniformed Services University
    of the Health Sciences
F. Edward Hebert School
    of Medicine
Bethesda, Maryland, U.S.A.
Chapter 7

Dagmar Ringe
Departments of Biochemistry and
    Chemistry
Brandeis University
Waltham, Massachusetts, U.S.A.
Chapter 1

Robin C. Sandlin
Department of Microbiology and
    Immunology
Uniformed Services University
    of the Health Sciences
F. Edward Hebert School
    of Medicine
Bethesda, Maryland, U.S.A.
Chapter 7

Nikolaus Schiering
Departments of Biochemistry and
    Chemistry
Brandeis University
Waltham, Massachusetts, U.S.A.
Chapter 1

Zeljka Sokolovic
Theodor-Boveri-Institut für
    Biowissenschaften (Biozentrum)
Lehrstuhl für Mikrobiologie
Universität Würzburg
Würzburg, Germany
Chapter 8

Marie-Paule Sory
Microbial Pathogenesis Unit
International Institute of Cellular
    and Molecular Pathology (ICP)
    and Faculté de Médecine
Université Catholique de Louvain
Brussels, Belgium
Chapter 6

Xu Tao
Evans Department of Clinical
    Research and Department
    of Medicine
Boston University Medical Center
    Hospital
Boston, Massachusetts, U.S.A.
Chapter 1

Hui-Yan Zeng
Evans Department of Clinical
    Research and Department
    of Medicine
Boston University Medical Center
    Hospital
Boston, Massachusetts, U.S.A.
Chapter 1

Bacteria can live in diverse and defined environments. Nutrient and toxin levels, temperature, osmolarity, acidity, population, and many other conditions can change rapidly and unexpectedly. In order to survive, cells must constantly monitor their structure and physiology, accordingly. These adaptive behaviors are established by using environmental sensors and response regulators. These communication modules regulate a wide variety of signals including host detection and invasion, cell cycle, metabolite utilization, starvation and many others. The purpose of this volume is to give an overview of the various systems and to introduce recent advances in understanding selected systems of pathogenic bacteria. The field of signal transduction systems is quite wide and rapidly growing, including eukaryotic systems. Therefore, a detailed analysis of all systems, although useful, is not here approachable. The hope is that by understanding the common denominators and differences of molecular and physiological mechanisms of various pathogens we may acquire a better knowledge of the environmental host response.

Dr. Rino Rappuoli
Dr. Vincenzo Scarlato
Dr. Beatrice Aricò

AN OVERVIEW OF BACTERIAL SIGNAL TRANSDUCTION

Rino Rappuoli, Vincenzo Scarlato,
Beatrice Aricò and Naomi Balaban

1. INTRODUCTION

The ability to sense and respond to the signals deriving from either the environment or other living organisms, is one of most important features of life. Pathogenic bacteria, like all living organisms, have developed efficient systems to scout the surroundings and adapt their life according to the signals that they can sense. The signals sensed by bacteria can be divided into three main categories: those deriving from the environment, those deriving from other organisms and those deriving from other bacteria (population signals). The signals perceived from the environment can be of physical or chemical nature, such as temperature, osmolarity, pH, light, CO_2, ammonia, oxygen, metals, nutrients, etc. The signals deriving from other living organisms may be either diffusable molecules such as chemoattractants, or signals that derive from direct contact with the organism. The signals deriving from other bacteria are usually diffusable molecules produced by the bacteria themselves which accumulate in the medium and increase in concentration with the bacterial cell density. This book contains nine chapters describing the regulatory systems of bacterial virulence that have been best characterized at the molecular level. The chapters go from the iron-mediated regulation of diphtheria toxin production that is the oldest report of environmental regulation of virulence, described by Pappenheimer and Johnson in 1936,[1] to the most recent reports on autocrine regulation of virulence expression in *Staphylococcus aureus* and *Pseudomonas aeruginosa*. In the following pages, we would like to give a general overview of the different systems, in an attempt to point out the mechanisms that are common to most bacteria. A summary of signals perceived by bacteria is shown in Figure I.1.

Signal Transduction and Bacterial Virulence, edited by Rino Rappuoli,
Vincenzo Scarlato and Beatrice Aricò. © 1995 R.G. Landes Company.

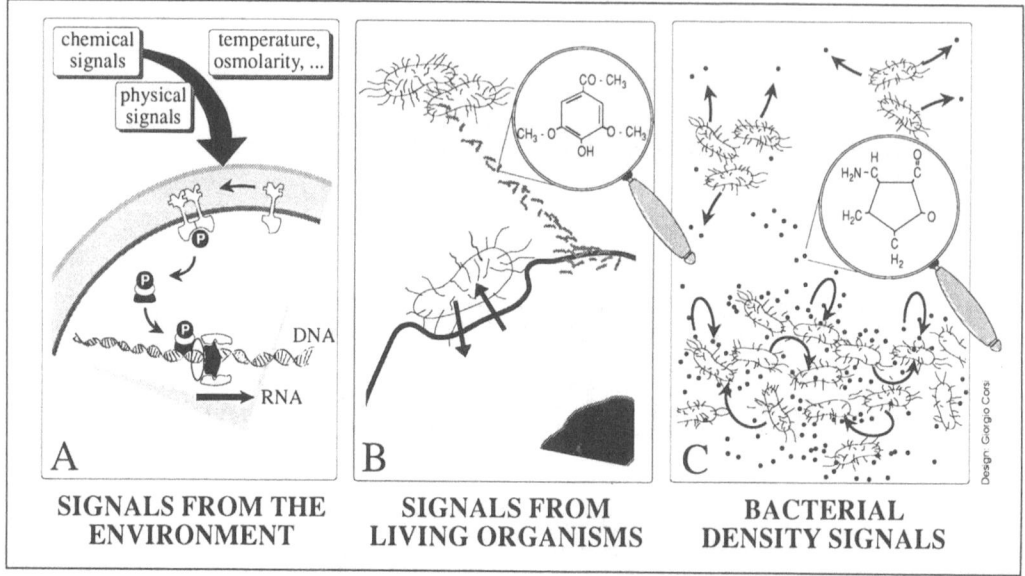

| SIGNALS FROM THE ENVIRONMENT | SIGNALS FROM LIVING ORGANISMS | BACTERIAL DENSITY SIGNALS |

Fig. I.1. Schematic representation of signals perceived by bacteria from (A) external signals, (B) other living organisms (Acetosyringone structure is magnified), (C) other bacteria (Homoserine laetone structure is modified).

2. SIGNALS DERIVING FROM THE ENVIRONMENT

Bacteria sense environmental signals such as temperature, osmotic pressure, phosphate, calcium, nutrients etc., and respond to the environmental changes by modifying gene expression in order to produce those molecules that are necessary to survive and multiply under each different condition.[2] Signals are usually detected by specific receptors that are located on the bacterial surface and belong to the two component family of regulatory systems (Fig. I.1A).[3] These are made by a transmembrane receptor that senses the signal and autophosphorylates a histidine residue present in the cytoplasmic portion of the molecule. The phosphate is then transferred to the second component of the system that is a cytoplasmic protein that once phosphorylated, changes conformation and activates gene transcription.[4,5] The two-component regulatory mechanism is probably the most widespread regulatory system in bacteria, and it is present also in eukaryotic organisms.[6-8] Each bacterium contains multiple couples of the two-component systems which respond

to different signals. In this book, the two-component system is shown to regulate virulence in *Bordetella*, *Pseudomonas*, *Salmonella*, *Vibrio cholerae*, *Shigella* and *Yersinia*, in response to temperature, osmolarity, chemical compounds, etc.

In rare cases, environmental signals are not transduced through the two-component systems. One example of this is the iron regulation of diphtheria toxin synthesis. In this case, the iron is taken up by specific transport systems, and once in the cytoplasm, binds directly to and activates the repressor of transcription.

3. SIGNALS DERIVING FROM OTHER LIVING ORGANISMS

Classical signals deriving from other living organisms, known to influence bacterial gene expression, are the phenolic compounds and monosaccharides produced by the wound roots of dicotyledonous plants that in *Agrobacterium tumefaciens* turn on the expression of the genes required for the transfer of the Ti plasmid DNA to the plant cells (Fig. I.1B).[9] Like the environmental signals, the signals involved in this system are transduced by a two-component system. Chemoattractants and repellents are other examples of diffusable molecules released into the environment by living organisms that may modify gene expression in bacteria.

Recently, great attention has been dedicated to the communication deriving from the direct contact between bacteria and different living organisms. These interactions occur typically during infection, when bacteria attach to or invade eukaryotic cells, and are often referred to as crosstalk between bacteria and the host.[10,11] The host-pathogen interaction can result either in de novo protein synthesis, or in the assembly and secretion of proteins that are already present. This book contains elegant examples of both systems. In *Yersinia*, the surface exposed YopN protein detects the signal that follows bacterial adhesion to the eukaryotic cell and induces the expression of a number of proteins of the Yop system that provide resistance to phagocytosis by macrophages. The most relevant in this system is the YopE protein, that following de novo synthesis, is transferred into the cytoplasm of the eukaryotic cells where it induces disruption of the actin microfilament network.[12,13] Another example of de novo protein synthesis is provided by the interaction of enteropathogenic *Escherichia coli* (EPEC) with epithelial cells. This contact induces the synthesis of bundle expressing pili that create a network of fibers that bind together individual organisms.[14,15]

In other cases, the contact of bacteria with eukaryotic cells does not result in the synthesis of new proteins. This is the case of *Salmonella* and *Shigella*, two intracellular pathogens that following the contact require a very rapid response that induces the intracellular translocation of the bacteria. In *Shigella*, the IpaB, IpaC and IpaD invasins that are required for entry in eukaryotic cells, accumulate in the cytoplasm and are released into the environment following contact with eukaryotic cells. The release of the proteins requires transient interaction of the IpaB and IpaD with the Mxi-Spa secretory

apparatus that is induced following a signal deriving from the host cell or serum.[16] In *Salmonella typhimurium*, contact with eukaryotic cells induces the formation of appendages on the surface of the bacterium which do not require de novo protein synthesis and disappear once the bacteria have initiated the internalization process.[11,17] On the other hand, eukaryotic cells also receive signals from the interacting bacteria and respond with a cascade of events leading to the formation of cell membrane ruffles and bacteria uptake.[11]

4. CELL DENSITY SIGNALS

Bacteria are not merely passive responders to environmental changes, but control their own regulatory systems. During growth, they continuously produce and secrete diffusable signals which enable them to signal their presence to other bacteria, to sense the presence of their neighbors, and to monitor their own population density (Fig. I.1C). These signal molecules induce autoregulatory systems and allow the bacterium to exist at a multicellular level, sensing the density of its local population and enabling it to act in concert. Some bacteria produce autoinducers that regulate only their own specific global regulons necessary for their survival. Other bacteria exhibit mutualism in regulating to different degrees global regulons of various bacterial species, allowing for example an orchestrated attack on the host. Conversely, other bacteria produce molecules that are toxic to other related bacterial populations, thus eliminating some of their competitors.

The more common family of autoinducers includes the various analogs of N-acyl homoserine lactones (HSLs). These are metabolites synthesized from intermediates of threonine biosynthesis which are released into the medium at high cell density, when nutrients become limiting, and may represent a general signal of starvation.[18] In *Vibrio fischeri*, where it has been best characterized,[19] the cell membrane is permeable to the HSL that accumulates into the medium and, at high cell densities, binds to LuxR, a 250 amino acid membrane bound transcriptional activator. Once activated, the LuxR binds 40 base pairs upstream from the transcriptional starting point of the *luxICDABEG* operon and induces transcription. The genes *DABEG* of this operon are involved in light production, while the *luxI* gene encodes a 193 amino acid protein that is the autoinducer synthetase.

These autoinducers have recently been described as the bacterial Esperanto,[23] universal language among Gram-negative bacteria. HSLs have been shown to control diverse range of cell density-dependent factors, including bioluminescence in *Vibrio fisheri* and *Vibrio harveyi*,[19] the control of antibiotic and exoenzyme production in *Erwinia carotovora*, Ti-plasmid conjugation in *Agrobacterium tumefaciens*, elastase production in *Pseudomonas aeruginosa*.[20] This last case, is a beautiful example showing how bacteria can use signals coming from the environment and cell density signals to coordinate the expression of their virulence factors, to achieve a concentrated attack to the host. HSLs can be considered a unique family of response regulators that use homoserine lactones rather than phosphorylation as their signals.

In addition to the HSLs which represent a general mechanism of bacterial communication, bacteria also have specific cell-density autoinducers that accumulate in culture medium during growth and exert their effect only when a threshold concentration of the autonducer is reached. The specific autoinducers that have so far been described include the production of a peptide pheromone that is required for the genetic competence and for efficient sporulation in *Bacillus subtilis*,[21] and the protein RAP which is required for virulence in *Staphylococcus aureus*.[22] It is easy to predict that during the next few years, many new examples of the autocrine regulation of virulence in pathogenic bacteria will be described.

REFERENCES

1. Pappenheimer AM, Johnson SJ. Studies on diphtheria toxin production. I: the effect of iron and copper. Br J Exp Pathol 1936; 17:335-341.
2. Miller JF, Mekalanos JJ, Falkow S. Coordinate regulation and sensory transduction in the control of bacterial virulence. Science 1989; 243:916-922.
3. Ronson CW, Nixon BT, Ausubel FM. Conserved domains in bacterial regulatory proteins that respond to environmental stimuli. Cell 1987; 49:579-581.
4. Gross R, Aricò B, Rappuoli R. Families of bacterial signal-transducing proteins. Mol Microbiol 1989; 3:1661-1667.
5. Alex LA, Simon MI. Protein histidine kinases and signal transduction in prokaryotes and eukaryotes. TIG 1994; 10:133.
6. Chang C, Kwok SF, Bleecker AB et al. Arabidopsis ethylene-response gene ETR1: Similarity of product to two-component regulators. Science 1993; 262:539-544.
7. Ota IM, Varshavsky A. A yeast protein similar to bacterial two-component regulators. Science 1993; 262:566-569.
8. Maeda T, Wurgler-Murphy SM, Saito H. A two-component system that regulates an osmosensing MAP kinase cascade in yeast. Nature 1994; 369:242-244.
9. Winans SC. Two-way chemical signaling in *Agrobacterium*-plant interactions. Microbiology Rev 1992; 56:12-31.
10. Wick MJ, Madara JL, Fields BN et al. Molecular cross talk between epithelial cells and pathogenic microorganisms. Cell 1991; 67:651-659.
11. Bliska JB, Galan JE, Falkow S. Signal transduction in the mammalian cell during bacterial attachment and entry. Cell 1993; 73:903-920.
12. Forsberg A, Rosqvist R, Wolf-Watz H. Regulation and polarized transfer of the *Yersinia* outer proteins (Yops) involved in antiphagecytosis. Trends in Microbiology. 1994; 2:14-19.
13. Rosqvist R, Magnusson KE, Wolf-Watz H. Target cell contact triggers expression

and polarized transfer of *Yersinia* YopE cytotoxin into mammalian cells. EMBO J 1994; 13:964-972.

14. Vuopio-Varkila J, Schoolnik GK. Localized adherence by enteropathogenic *Escherichia coli* as an inducible phenotype associated with the expression of new outer membrane proteins. J Exp Med 1991; 174:1167-1177.

15. Giron JA, Suk Yue Ho A, Schoolnik GK. An inducible bundle-forming pilus of enteropathogenic *Escherichia coli*. Science 1991; 254:710-713.

16. Menard R, Sansonetti P, Parsot C. The secretion of the *Shigella flexneri* Ipa invasins is activated by epithelial cells and controlled by Ipa B and Ipa D. EMBO J 1994; 13:5293-5302.

17. Galan JE. Salmonella entry into mammalian cells: Different yet converging signal transduction pathways? Trends in Cell Biol 1994; 4:196-199.

18. Huisman GW, Kolter R. Sensing starvation: A homoserine lactone-dependent signaling pathway in *Escherichia coli*. Science 1994; 265:537.

19. Claiborne Fuqua W, Winans SC, Greenberg EP. Quorom sensing in bacteria: the LuxR-LuxI family of cell density-responsive transcriptional regulators. J Bacteriol 1994; 176:269-275.

20. Passador L, Cook JM, Gambello MJ et al. Expression of *Psuedomonas aeruginosa* virulence genes requires cell-to-cell communication. Science 1993; 260:1127.

21. Magnuson R, Solomon J, Grossman AD. Biochemical and genetic characterization of a competence pheromone from *B. subtilis*. Cell 1994; 77:207-216.

22. Balaban N, Novick RP. Autocrine regulation of toxin synthesis by *Staphylococcus aureus*. Proc Natl Acad Sci USA 1995; 92:1619-1693.

23. Swift S, Bainton NJ, Winson MK. Gram-negative bacterial communication by N-acyl homoserine lactones: a universal language.? Trends Microbiol 1994; 2:193-198.

SIGNAL TRANSDUCTION AND IRON-MEDIATED REGULATION OF VIRULENCE FACTORS

Xu Tao, Nikolaus Schiering, Hui-Yan Zeng,
Dagmar Ringe and John R. Murphy

1. INTRODUCTION

The coordinate control of virulence determinants in pathogenic microorganisms is largely based upon the ability of these microbes to rapidly adapt to the environment presented by their host. Essentially all microbial pathogens have evolved specific mechanisms for the assimilation of sufficient concentrations of iron from their environment to support growth. It is remarkable that sensing the available concentration of iron provides not only a signal for siderophore expression, but also a regulatory signal for the expression of a wide variety of bacterial toxins and other virulence factors (reviewed in ref. 1). In *Escherichia coli* it is widely known that the coordinate regulation of iron-sensitive genes is mediated by Fur (ferric uptake regulator). Once it is activated by iron, Fur has been shown to function as a global regulatory element controlling the expression of regulons that are distributed throughout the *E. coli* chromosome. Moreover, it is now clear that the iron-mediated regulation of virulence genes in a number of Gram-negative pathogens is also coordinated by a family of closely related proteins that are homologous to Fur.

We have known for almost 60 years that the production of diphtheria toxin, the primary virulence factor of toxigenic strains of *Corynebacterium diphtheriae*, is extremely sensitive to the concentration of iron in the bacterial culture medium. During the past several years, the understanding of the molecular mechanisms involved in the iron-mediated regulation of diphtheria toxin expression has increased dramatically. The molecular cloning and subsequent characterization of

Signal Transduction and Bacterial Virulence, edited by Rino Rappuoli,
Vincenzo Scarlato and Beatrice Aricò. © 1995 R.G. Landes Company.

the iron-dependent diphtheria *tox* repressor, DtxR, from *C. diphtheriae* has allowed for the demonstration that this regulatory protein governs the expression of at least those genes encoding toxin and siderophore production. Thus, DtxR appears to be the *C. diphtheriae* equivalent of Fur. While DtxR and Fur appear to be functionally equivalent, their target DNA binding sequences differ, and as a result they are incapable of cross-regulation of cistrons under their respective control.[2] Schmitt and Holmes[3] have shown that the introduction of *dtxR* into the iron insensitive mutant strain *C. diphtheriae* C7hm723(β) was able to restore full repression of siderophore production and partial repression of toxin production in a high iron environment.

While iron-mediated regulation of bacterial toxin and virulence gene expression has been well documented, the binding of iron by apo-DtxR, as well as the steps involved in the activation of this metal-ion dependent regulatory element are only partially understood. It has been shown that DtxR, in the presence of heavy metal ions, specifically binds to and protects the diphtheria *tox* operator from DNase I digestion.[4,5] The stoichiometry of metal ion binding, the amino acid residues involved in forming the metal ion coordination sites, and the details of interaction between activated DtxR and its DNA target sequence(s) are beginning to be resolved. Since several excellent reviews on iron-mediated regulation of the expression of virulence factors have been published, this review will focus on our current understanding of the structure function relationships of DtxR. We shall also present a working model of the molecular events involved in the metal ion-activation of DtxR which lead to its binding to the *tox* operator and the subsequent repression of diphtheria *tox* expression.

2. IRON IS AN ESSENTIAL NUTRIENT

Even though iron is one of the most abundant elements in the earth's crust,

because of its poor solubility it is not readily available to living organisms. In aqueous solution, iron may exist in either ferrous (Fe^{2+}) or ferric (Fe^{3+}) forms. Under aerobic conditions ferrous iron rapidly becomes oxidized to the ferric form which is less soluble and, as a result biologically available concentrations may be as low as 10^{-18} M at neutral pH. Since this level is far below the concentration required by living organisms, highly efficient iron transport systems have evolved to facilitate the uptake of iron from the environment.

In prokaryotes and lower eukaryotes environmental iron is scavenged from the environment and transported into the cytosol by a group of low-molecular-weight iron-chelators, the siderophores (from the Greek "iron bearers"). At present, more than 200 different natural siderophores have been identified.[6] Although siderophores display a considerable structural diversity, they all form six coordinate octahedral complexes with ferric iron, and they all have an extremely high affinity for ferric iron with stability constants ranging from 10^{23} to 10^{52}. When the concentration of iron in the environment is low, the genes encoding many of the siderophores and their receptors are expressed de novo.

In order to survive as parasites, bacterial pathogens must respond to diverse environments, ranging from their native habitat to their sensitive host, and coordinate the expression of sets of genes which allow them to survive. Since iron is an essential element for both the bacterial pathogen and its host, competition for this nutrient is an essential component of the infectious process. Due to the presence of mammalian iron binding proteins the concentration of iron available to an invading bacterial pathogen is extremely limited. Given the fact that an invading pathogen must colonize and replicate in situ, it is not surprising that the regulation of many virulence determinants (e.g., colonization factors, hemolysins, toxins) is controlled by iron binding proteins.

3. IRON-MEDIATED REGULATION OF DIPHTHERIA *tox* EXPRESSION

In 1936, Pappenheimer and Johnson[7] demonstrated that the addition of iron salts to the growth medium of toxigenic *Corynebacterium diphtheriae* resulted in the inhibition of diphtheria toxin production.[7] Subsequently, it was found that diphtheria toxin was expressed at maximal rates only during the decline phase of the bacterial growth cycle when iron became the growth rate limiting substrate.[8] Uchida et al[9] demonstrated that the diphtheria toxin structural gene, *tox*, was carried on the genome of corynebacteriophage β. However, it also was apparent that the expression of *tox* was dependent upon the physiologic state of the *C. diphtheriae* bacterial host. As long as toxigenic strains of *C. diphtheriae* were grown under iron-limiting conditions the *tox* gene could be expressed from the prophage genome,[10] from nonintegrated repressed genome,[11] and from replicating exogenotes during vegetative phage growth.[12] However, the addition of iron to the culture medium of iron-limited *C. diphtheriae* resulted in the rapid repression of toxin synthesis.[13,14]

The demonstration that *tox* was encoded on the corynebacteriophage genome and that the regulation of *tox* expression was apparently mediated by the physiologic state of *C. diphtheriae* raised the interesting question of whether the *tox* gene was controlled by a corynebacteriophage gene product or a corynebacterial determined factor. In order to separate corynebacteriophage from corynebacterial factors in the regulatory process, Murphy et al[15] used S-30 extracts of *E. coli* to direct the synthesis of diphtheria *tox* gene products in vitro. When β-phage DNA was added to the *E. coli* S-30 system, diphtheria toxin as well as other phage proteins were synthesized. These workers found that the synthesis of toxin was not affected by the addition of iron to this in vitro heterologous coupled transcription/translation system. In contrast, even though other β-phage proteins were produced, diphtheria toxin could not be synthesized in S-30 extracts prepared from a nonlysogenic strain of *C. diphtheriae* and supplemented with iron. Most important, however, the addition of small amounts of the *C. diphtheriae* S-30 extract to the *E. coli* S-30 system programmed with β-phage DNA resulted in the specific inhibition of toxin synthesis. These results led to the hypothesis that *C. diphtheriae*, regardless of its lysogenic state, contained a factor which acted as a negative controlling element in the regulation of toxin production.

Additional evidence in support of this hypothesis came from Kanei et al[16] who described the isolation of several mutants of *C. diphtheriae* that constitutively expressed diphtheria toxin irrespective of the concentration of iron added to the growth medium. Corynebacteriophage β released from these mutants were used to infect nonlysogenic strains of *C. diphtheriae*, and toxin production in the newly formed lysogens was found to be under wild type control. These experiments demonstrated that the mutation which gave rise to constitutive *tox* expression was carried on the *C. diphtheriae* chromosome, and possibly in the gene encoding the iron-dependent *tox* regulatory element.

It was also shown that corynebacteriophage carried a cis-element responsible for the regulation of diphtheria *tox* expression. Murphy et al[14] isolated a β-phage mutant, β$_{ct1}$, which converts the wild-type C7(−) strain to an iron-insensitive Tox$^+$ phenotype. Double lysogens of C7(β*tox*-45/β$_{ct1}$) were found to produce both CRM45 (a nontoxic fragment of diphtheria toxin resulting from a nonsense mutation in the *tox* gene) and native diphtheria toxin under iron-limiting conditions; whereas, only toxin was produced in the presence of inhibitory concentrations of iron. The simplest explanation of these results was that the diphtheria *tox* gene is under control of a cis-element, and that a mutation in this element results in constitutive toxin production.

Based on these observations, Murphy and Bacha[17] proposed a model of *tox* regulation in which both a *C. diphtheriae* encoded determinant and corynebacteriophage element were involved in the regulation of toxin expression. The model postulated that *C. diphtheriae* carried a gene encoding for an aporepressor which in the presence of iron formed an active complex that specifically bound to the *tox* operator and blocked transcription (Fig. 1.1). Subsequently, Welkos and Holmes[18] were also successful in the isolation of corynebacteriophage mutants that were cis-dominant for *tox* expression. Genetic mapping experiments indicated that these mutations were closely linked to the 5'-end of the diphtheria *tox* structural gene,[19] and more recently DNA sequence analysis demonstrated that these mutations were, in fact, in the *tox* operator.[20] While results from both the biochemical and genetic analyses of diphtheria *tox* gene regulation were consistent with portions of the model proposed by Murphy and Bacha,[17] rigorous proof of this hypothesis could only be made by experiments conducted at the molecular level.

4. DIPHTHERIA *TOX* REGULATORY SEQUENCES

The DNA sequence of the diphtheria *tox* promoter operator region is shown in Figure 1.2. In addition to the "-35" and "-10" elements of the promoter,[21,22] a 9 base pair (bp) interrupted palindromic sequence representing the putative *tox* operator was identified.[23-25] Fourel et al[26] were the first to demonstrate that a factor(s) in crude *C. diphtheriae* extracts could specifically bind to a *tox* operator probe and protect it from DNase I digestion. Since protection of the probe was dependent upon the addition of iron to the reaction mixture, it was suggested that the binding factor(s) was the diphtheria toxin repressor. However, since only crude bacterial extracts were used in this study, it was difficult to ascertain whether the protective factor(s) was, in fact, the diphtheria *tox* repressor. The formal proof of the existence of an iron-dependent diphtheria *tox* repressor came only with the molecular cloning of this genetic element and its subsequent characterization.

5. MOLECULAR CLONING AND CHARACTERIZATION OF THE DIPHTHERIA *TOX* REPRESSOR, DTXR

Boyd et al[2] and Schmitt and Holmes[3] described the molecular cloning of an iron-dependent regulatory element, *dtxR* (diphtheria toxin regulatory protein), from ge-

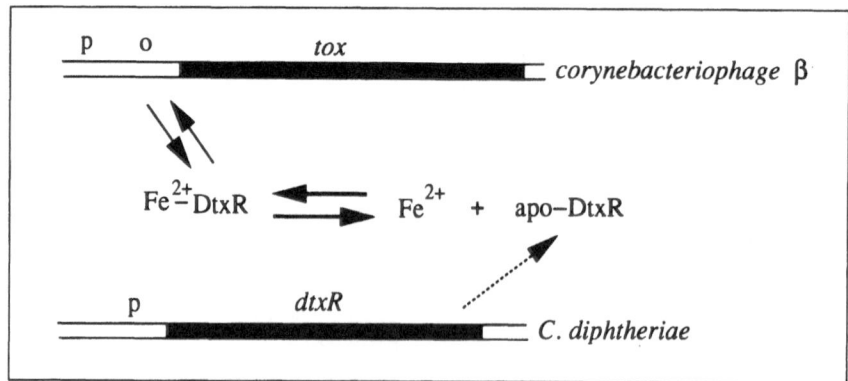

Fig. 1.1. Model of diphtheria tox regulation by the diphtheria tox repressor (DtxR). The structural gene for diphtheria toxin, tox, in carried on the genome of corynebacteriophage β tox+; whereas, the structural gene encoding DtxR is carried on the Corynebacterium diphtheriae genome. Apo-DtxR is activated in the presence of Fe^{2+} and binds to the tox operator thereby preventing transcription. As iron becomes the growth rate limiting substrate for C. diphtheriae, Fe^{2+} dissociates from DtxR and it is converted to apo-DtxR and the tox gene becomes derepressed. Modified from Murphy and Bacha.[17]

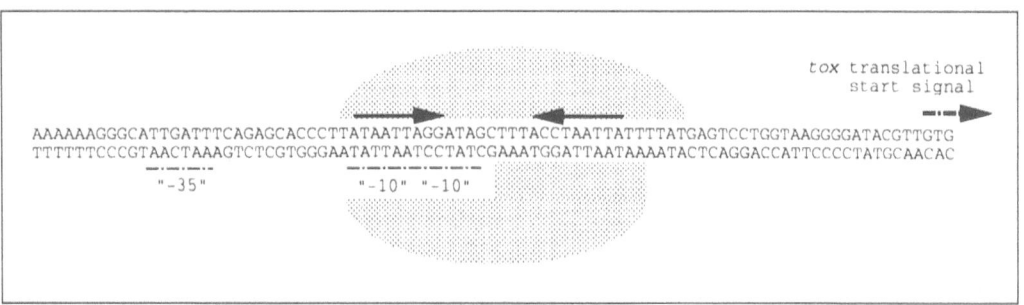

Fig. 1.2. Deoxyribonucleic acid sequence of the diphtheria tox promoter / operator region upstream of the diphtheria tox structural gene. The positions of diphtheria tox translational start site (broken arrow), the "–35" and "–10" regions of the tox promoter, and the 9-base pair palindromic sequence of the tox operator (solid arrows) are shown. The nucleotide sequence covered by the shaded areas indicate the region protected from DNase I digestion in the presence of activated DtxR.

nomic libraries of nontoxigenic, nonlysogenic *C. diphtheriae*. In recombinant *E. coli*, DtxR was shown to repress the expression of β-galactosidase from a *toxPO/lacZ* transcriptional fusion in an iron-dependent fashion.[2] Importantly, the report of Schmitt and Holmes[3] and Tai et al[27] also demonstrated that *DtxR* could restore iron-mediated regulation of diphtheria *tox* and siderophore gene expression in the iron-insensitive C7hm723(βtox+) strain of *C. diphtheriae*. These experiments provided the first definitive proof that DtxR was, in fact, a global regulatory element involved in the control of iron sensitive genes in *C. diphtheriae*.

DNA sequence analysis indicates that *dtxR* encodes a 25,316 molecular weight protein. More recently, the *dtxR* alleles from PW8(–), 1030(–) and C7hm723(–) strains of *C. diphtheriae* were also cloned and sequenced.[28,29] It is interesting to note that the *dtxR* allele from the PW8(–) strain was found to be identical to that from C7(–); whereas, DtxR from the *C. diphtheriae* (belfanti) 1030(–) strain was found to carry six amino acid substitutions in the C-terminal region of the protein. These substitutions appear to be silent mutations and do not affect the iron-dependent regulatory activity of DtxR(1030).

DNA sequence analysis of the *dtxR* allele from the nitrosoguanidine-induced mutant strain C7hm723(–) revealed a single G to A transition mutation resulting in the substitution of Arg47 to His

(R47H).[28,29] DtxR(R47H) only weakly regulates the expression of β-galactosidase from the *toxPO/lacZ* transcriptional fusion in recombinant *E. coli*.[28] Based upon the analysis of DtxR mutants[30] and identification of the helix-turn-helix DNA binding motif in the electron density map of DtxR[31] Arg47 is positioned in the DNA binding domain and is likely to be directly involved in the interaction between the repressor and the diphtheria *tox* operator.

6. ACTIVATION OF DTxR REQUIRES TRANSITION METAL IONS

Using purified DtxR, Tao et al[32] demonstrated by gel mobility shift experiments that the binding of DtxR to a [^{32}P]-labeled diphtheria *tox* operator probe required the presence of divalent heavy metal ions. The specificity of the interaction between metal ion-activated DtxR and the *tox* operator was demonstrated by showing that the addition of either excess cold probe, anti-DtxR antisera, or the chelator 2,2'-dipyridyl to the reaction mixture blocked binding to the labeled probe.

Both Tao and Murphy[4] and Schmitt and Holmes[5] have demonstrated that heavy metal ion-activated DtxR is able to protect the *tox* operator from DNase I digestion. In the presence of either Co^{2+}, Fe^{2+}, or Mn^{2+}, DtxR was found to protect 33-nt and 27-nt regions on the coding and noncoding strands, respectively

(Fig. 1.2). In these instances, the DtxR footprint on the noncoding strand directly coincides with the 9-bp interrupted palindromic sequence. The activation of DtxR with Ni^{2+} results in a slightly smaller footprint, 31-nt and 26-nt on the coding and noncoding strand. Schmitt and Holmes[5] have reported almost identical results, and in addition have shown that Cd^{2+} is also able to activate DtxR. In contrast to the other divalent heavy metal ions, the addition of Zn^{2+} to a concentration of ≥ 100 µM was only able to partially activate DtxR; whereas, Cu^{2+} failed to activate this regulatory element in vitro.[4,5] It is of interest to note that the heavy metal ions used in vitro to activate DtxR were also found to inhibit the production of diphtheria toxin by toxigenic strains of *C. diphtheriae* in vivo.[33]

7. DtxR METAL ION BINDING ACTIVATION SITE

The DNA sequence analysis of DtxR[2,3] revealed that there was a single cysteine residue at position 102. Since purified DtxR readily forms disulfide-linked dimers under oxidizing conditions, Cys102 appeared to be exposed on the surface of this regulatory protein. Moreover, in dimeric form DtxR fails to bind to the *tox* operator probe even in the presence of elevated concentrations of divalent heavy metal ions. In order to determine whether Cys102 was positioned in the metal-binding activation site, Tao and Murphy[34] used saturation site-directed mutagenesis of codon-102. Mutant *dtxR* alleles carrying each of the 20 amino acids at this position were isolated and characterized. Each of the cloned *dtxR* alleles were separately introduced into *E. coli* DH5α:λRS45toxPO/lacZ, a reporter strain which carries a single genomic copy of a transcriptional fusion between the diphtheria *tox* regulatory region and *lacZ*. From all of the *dtxR* alleles studied, only the wild type (TGC) and 2 mutants were found to encode an active metal ion dependent regulatory element. The two mutant *dtxR* alleles carried either the alternate codon for cysteine

(TGT) or an aspartic acid codon (GAC) at position 102.

In order to demonstrate rigorously that DtxR(C102D)-mediated repression of *lacZ* was iron dependent, Tao and Murphy[34] recloned the *dtxR*-102D allele into pRS551*toxPO/lacZ*. The results obtained from quantitative measurement of β-galactosidase produced in strains which carried either the wild type *dtxR* or the *dtxR*-102D allele demonstrated that both DtxR and DtxR(C102D) functioned as iron-dependent regulatory elements. Even though the heavy metal ion-dependent regulatory activity of DtxR(C102D) was not as sensitive to the concentration of iron as wild type DtxR, the mutant was able to control β-galactosidase expression from the *toxPO/lacZ* transcriptional fusion in an iron-dependent manner. Tao and Murphy[34] also demonstrated by gel-mobility shift experiments that the interaction between DtxR(C102D) and the *tox* operator is both specific and dependent upon the presence of heavy metal ions.

The observations that only Asp was able to substitute for Cys at position 102, and that DtxR(C102D) was also an iron-dependent regulatory element provide strong support for the argument that Cys102 plays an important role in DtxR-mediated regulatory activity. Cysteine and aspartic acid have different physical properties, however, both of these amino acids have been shown to serve as metal binding residues in many metalloproteins.[35] These results led to the conclusion that Cys102 in DtxR is positioned in the heavy metal ion activation site and plays an essential role in the action of this regulatory protein.

Recently, Wang et al[30] have described a series of mutations in *dtxR* that inactivate the regulatory activity of the repressor. In particular, the three mutations, E100K, W104Q and H106Y, which lead to a marked decrease in repressor function are consistent with the argument that the divalent heavy metal ion activation domain of DtxR is positioned in this region of the regulatory element.

8. CRYSTALLOGRAPHIC STUDIES WITH DTXR

Using highly purified DtxR, we have been successful in obtaining crystals which diffract to better than 3.0 Å.[31] The space group of the crystals is trigonal P3121 or its enantiomorph P3221 with $\underline{a} = \underline{b} = 64.2$ Å, $\underline{c} = 220.5$ Å, $\alpha = \beta = 90°$, $\gamma = 120°$. Two monomers of DtxR are contained within the asymmetric unit. To date, we have been able to position 173 of 226 amino acids in the resulting electron density map. The helix-turn-helix DNA binding motif and the metal ion-activation site have been positioned in the N-terminal portion of the crystal structure which is comprised of amino acids 3 through 136 (Fig. 1.3A). As shown in Figure 1.3B, the metal ion binding domain of DtxR is likely to be composed of two α-helices in which His79, Glu83, His98, Cys102, Glu105, and His106 form the potential coordination sites for the metal ion binding site (Schiering, Tao, Murphy, Ringe, unpublished data). While these residues could potentially form a hexahedral coordination site for Fe^{2+}, at this time we cannot rule out the possibility that other residues in DtxR play a role in the formation of the metal ion binding site. It should be pointed out that these observations are consistent with the finding by Wang et al[30] that the R77H mutation leads to a marked reduction in DtxR function, and that the mutations A72V, R84H, and D88N result in decreased repressor activity. The latter mutations are all positioned in the same α-helix as His79.

As shown in Fig. 1.3A, the helix-turn-helix DNA binding domain on each monomer of DtxR is positioned such that the α-helices that are likely to be directly involved in binding to DNA are positioned 32 Å apart in the unit cell. This spacing is close to the 34-36 Å spacing that is optimal for DNA binding proteins which interact with sequences in the major groove. While it is likely that Q43, R47 and R50 of DtxR form hydrogen bonds with specific nucleotides in the target palindrome, the X-ray crystal structure that

has been determined is of apo-DtxR and the final assignment of those amino acid residues which bind to DNA must await the solution of the structure of co-crystals between metal ion activated DtxR and it target sequence (Schiering N, Tao X, Murphy JR, Ringe D, unpublished data).

9. STOICHIOMETRY AND COOPERATIVITY OF METAL ION BINDING BY DTXR

In order to further characterize metal ion binding by DtxR, we have measured the binding of $[^{63}Ni^{2+}]$ to apo-DtxR(C102D) by equilibrium dialysis (Zeng H, Tao X, Murphy JR, unpublished data). Since DtxR rapidly forms inactive disulfide linked dimers upon dialysis, the DtxR(C102D) mutant was used in these experiments. As deduced from Figure 1.4, each molecule of apo-DtxR(C102) carries a single high affinity binding site for $[^{63}Ni^{2+}]$ with a K_d of 1.2×10^{-6} M. It should be noted that these experiments employed Ni^{2+} as the activation metal because of the well known instability of Fe^{2+} ions under aerobic conditions. Tao and Murphy[4] have shown that the activation of apo-DtxR by Ni^{2+} occurs at concentrations that are similar to that determined for Fe^{2+}. This study has confirmed and extended the observations of Wang et al[30] that monomeric DtxR has a single metal ion binding site.

We have also measured the quenching of intrinsic tryptophan fluorescence upon the addition of heavy metal ions to both apo-DtxR and apo-DtxR(C102D) (Zeng H, Tao X, Murphy JR, unpublished data). Since DtxR contains only one tryptophan residue at position 104 in the metal ion binding domain, the quenching of fluorescence at 330 nm is a direct measure of the environment surrounding the indole ring of this amino acid. Figure 1.5 shows that upon addition of Ni^{2+} there is a quenching of Trp104 fluorescence. These results demonstrate that, at least in the region of the activation domain, DtxR undergoes a conformational change upon metal ion binding. Moreover, the sigmoidal curve of quenching of intrinsic tryptophan fluorescence suggests that Ni^{2+} binding by DtxR is cooperative. Since monomeric DtxR has only one metal ion binding site, the cooperativity of metal ion binding suggests that during the activation process apo-DtxR is in a dimeric form. Accordingly, we propose that in solution and in the absence of heavy metal ions, monomeric apo-DtxR is in weak

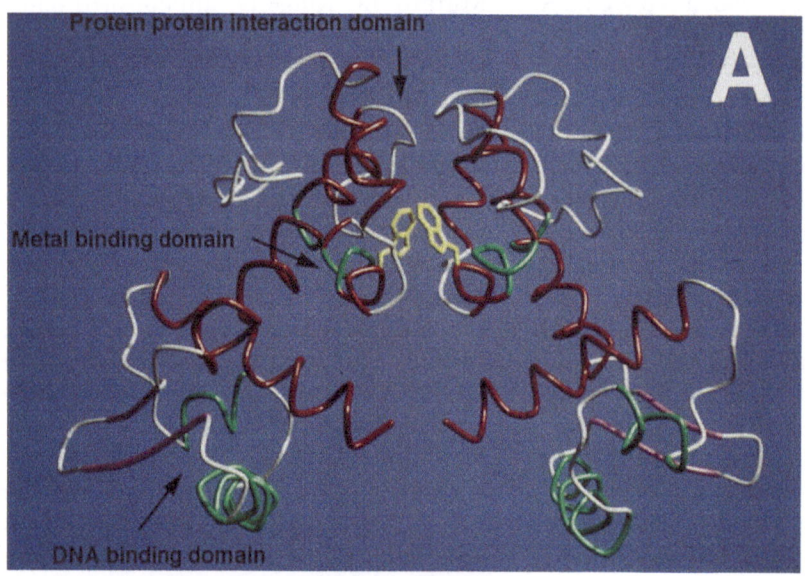

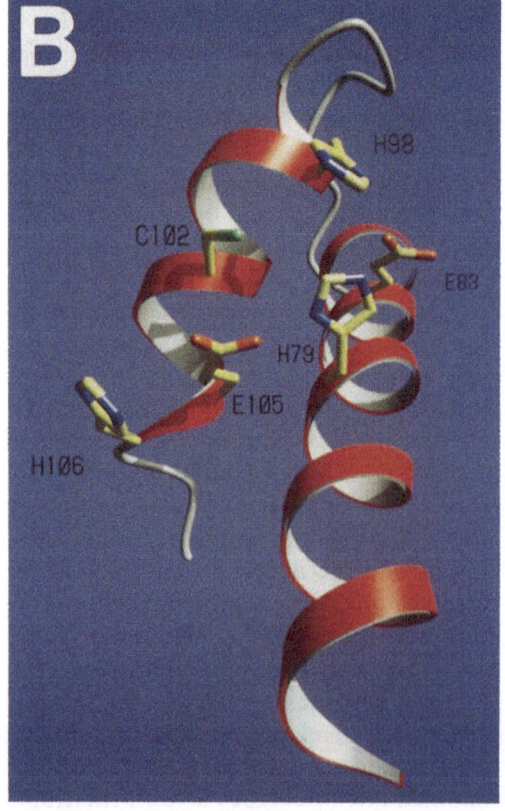

Fig. 1.3. (A) Preliminary X-ray crystal structure of the N-terminal region of apo-DtxR at 2.5 Å resolution. The X-ray structure shown is of amino acids 3 through 136 which contain the DNA binding, metal ion binding, and protein protein interaction domains. (B) X-ray crystal structure of the metal ion-activation site of DtxR. The structure is from amino acid 70 through 107 and shows those residues (His79, Glu83, His98, C102, Glu105, and His106) that are likely to play a role in the coordination of heavy metal ions.

equilibrium with a dimeric form. Further support for this hypothesis comes from experiments in which the quenching of intrinsic tryptophan fluorescence was examined. As discussed above, DtxR contains a single tryptophan residue at position 104 which is positioned in close proximity to the metal ion-activation domain. Since the side chains of Trp104 form part of the hydrophobic protein-protein inter-

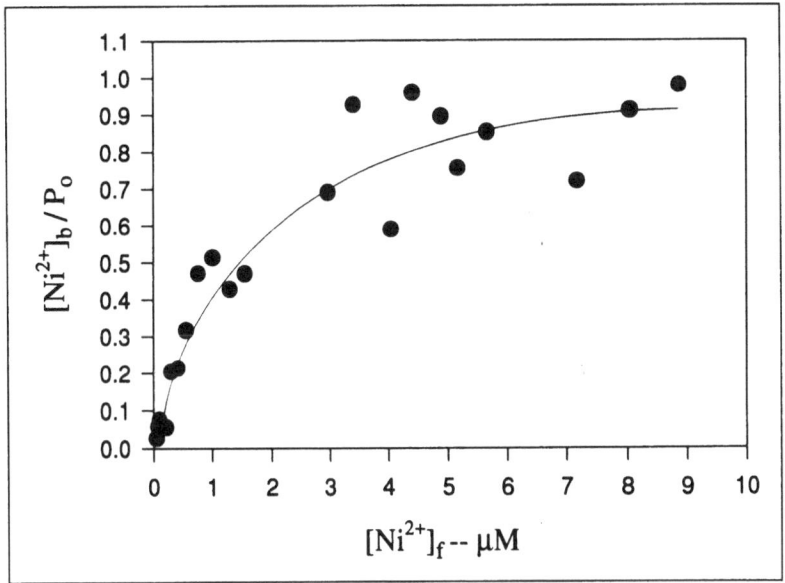

Fig. 1.4. Equilibrium dialysis of [^{63}Ni^{2+}] binding to apo-DtxR(C102D). The ordinate, [Ni^{2+}]$_b$/P$_o$, is the ratio of the concentration of DtxR(C102D) to which metal ions are bound versus the total concentration of monomeric DtxR(C102D).

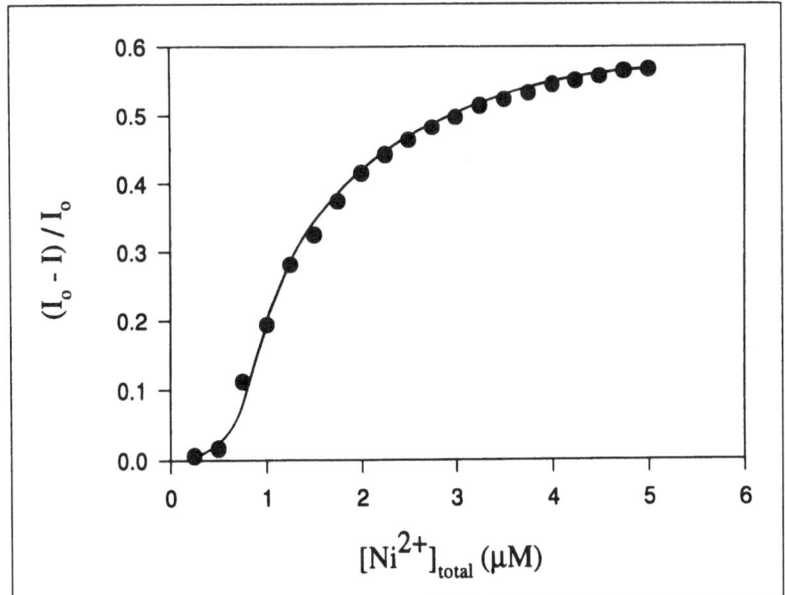

Fig. 1.5. Titration of the intrinsic tryptophan quenching of apo-DtxR upon the addition of Ni^{2+}. I$_o$, is the initial fluorescence corrected for volume; I, is the fluorescence after the addition of Ni^{2+}; [Ni^{2+}]$_t$, is the total concentration of nickel added to the reaction mixture.

action domain in the dimer (Fig. 1.3A), the binding of heavy metal ions by apo-DtxR would be anticipated to introduce a con-formational change that would alter the environment around Trp104. Since the binding of Ni^{2+} by apo-DtxR is cooperative, it is likely that binding of the first metal ion induces a conformational change that facilitates the binding of the second metal ion which, in turn, results in the stabilization of the dimeric complex and activation of functional DNA binding activity. At the present time, however, it is not known whether activated DtxR binds to target operator sequences as a dimer, tetramer, or as a higher order complex.

The hypothesis that apo-DtxR exists in solution as a monomer in weak equilibrium with a dimeric form has been recently strengthened. Tao, Zeng, and Murphy (unpublished data) have found that the addition of the inactive DNA binding domain deletion mutant DtxR(Δ1-47) to apo-DtxR results in the competitive inhibition of the formation of a functional activated DtxR. This study also adds support to the hypothesis that DtxR also carries a protein-protein interaction (PPI) domain whose function is to stabilize the dimeric form of the repressor. Indeed, preliminary cross-linking experiments with glutaraldehyde strongly suggest that monomeric apo-DtxR may be readily converted to a dimeric form. These observations are also consistent with the position and orientation of the two molecules of DtxR that are observed in the unit cell of the X-ray crystal structure of the repressor (Fig. 1.3A).

10. DtxR CONSENSUS BINDING SITE

Schmitt and Holmes[36] have recently described the molecular cloning and characterization of two iron responsive promoters from *C. diphtheriae* and have proposed a 19-bp DtxR consensus binding site. Independently, Tao and Murphy[37] have used the CASTing method[38] to identify the minimal essential nucleotide base

sequence for DtxR binding, and also have determined the DtxR consensus binding sequence.

Beginning with a pool of dsDNA in which 18-bp corresponding to the intervening and downstream portions of the interrupted palindromic sequence of the *tox* operator were randomized, Tao and Murphy[37] have selected in vitro a family of double stranded (ds) DNAs from a universe of 6.9 x 10^{10} potential binding sites in which the apparent DtxR binding affinity is equivalent to that of the wild type diphtheria *tox* operator. DNA sequence analysis of the in vitro affinity selected dsDNAs reveal that the consensus DtxR binding site is, in fact, smaller than the 27-bp native *tox* operator. The minimal essential DNA target site required for DtxR binding is a 19-bp sequence that forms a perfect palindrome around a central C or G. The consensus DtxR binding sequence as determined by in vitro affinity selection is as follows:

```
      A           C           A
5'-T.AGGTTAG.CTAACCT.A-3'
      T           G           T
```

Tao and Murphy[37] have also demonstrated that the in vitro affinity selected DtxR binding sites were able to function as iron-responsive operators in vivo. Six unique independently isolated DtxR binding sites were recloned into *toxPO/lacZ* transcriptional fusion vectors. The expression of β-galactosidase from the resulting recombinant strains was found to be regulated by DtxR and to vary by only 2- to 3-fold from the wild type *tox* promoter / operator *lacZ* fusion. It is remarkable that the DtxR consensus sequence determined by in vitro genetic methods is identical to that determined by the sequence analysis of two DtxR sensitive operators cloned from genomic libraries of *C. diphtheriae*.[36]

11. A WORKING MODEL OF IRON ACTIVATION OF APO-DtxR

Our current working model of the process by which apo-DtxR is activated by iron is shown in Figure 1.6. Experiments described above suggest that DtxR is composed of at least the following three functional domains: (i) DNA binding domain [**B**], (ii) a transition metal ion-activation domain [**A**], and (iii) a protein-protein interaction domain [**PPI**]. There are two lines of evidence which lead us to conclude that in solution monomeric apo-DtxR is in weak equilibrium with the dimeric form. The first is based on the observation that the binding of Ni^{2+} by apo-DtxR is cooperative; and the second is that there is a competitive inhibition of the formation of functional DtxR in the presence of DtxR(Δ1-47) only when the mutant is added to the mixture prior to the addition of metal ions. By analogy with other allosteric proteins, it is likely that the binding of the first metal ion induces a conformational change in DtxR that facilitates the binding of the second metal ion to a dimeric complex. The failure of DtxR(Δ1-47) to inhibit DtxR activation once it has bound metal ions, suggests that the equilibrium of the reaction leading to the activated dimeric form lies to the right. As shown in Figure 1.6, the process by which DtxR is activated most likely involves a series of conformational changes which lead to a form of the repressor which is able to bind to its target DNA and regulate the expression of iron-sensitive genes.

The model that we have proposed implies that the conversion of apo-DtxR to the active repressor is an extremely sensitive genetic switch. In vitro studies in which we have measured the quenching of intrinsic tryptophan fluorescence, demonstrate that apo-DtxR binds Fe^{2+} with a K_d of 5 x 10^{-7} M. It should be noted that this value is in close agreement with the observation that the addition of ≤ 1 x 10^{-6} M Fe^{2+} was sufficient to activate DtxR binding to a *toxPO* probe.[4] Thus, as the concentration of available iron in the *C. diphtheriae* cytosol falls to $\leq 10^{-6}$ M, iron would begin to dissociate from DtxR. We anticipate that the dissociation of iron from the active complex results in a conformational change in the repressor which, in turn, leads to the dissociation of the repressor from its target operators. Concomitantly, the conversion of DtxR to its apo-form would result in the derepression of iron-sensitive genes in *C. diphtheriae*. These genes would include those encoding for both siderophore and diphtheria toxin production.

12. IS THERE A DtxR FAMILY OF REGULATORY PROTEINS IN OTHER GRAM-POSITIVE EUBACTERIA?

Günter et al[39] have reported that a repressor-binding site in the promoter region of *desA* gene is responsible for iron-mediated gene regulation in Gram-positive *Streptomyces pilosus* and *Streptomyces lividans*. This dyad symmetrical element was found to be homologous to the native

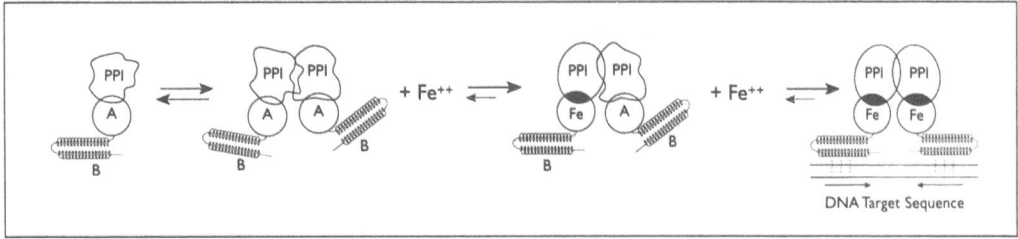

Fig. 1.6. Schematic representation of a working model of the Fe^{2+} activation of apo-DtxR. According to the model, monomeric apo-DtxR is in weak association with a homodimeric form of the repressor. The binding of Fe^{2+} to one of the monomers induces a conformational change in the structure which partially stabilizes the homodimer and results in facilitating the binding of the second Fe^{2+} ion to the complex. Upon binding of the second Fe^{2+} ion, the dimeric complex is stabilized and DtxR becomes functionally active and capable of binding with high affinity to its target DNA sequence.

diphtheria *tox* operator. The *desA* gene encodes lysine decarboxylase, which catalyzes the first reaction of siderophore desferrioxamine B synthesis, and is induced by iron deficiency.[40] Although the negative controlling element for *desA* has not been identified in *Streptomyces* sp., the remarkable similarity of the regulatory sequence to the consensus DtxR binding site suggests that a DtxR-like regulatory protein(s) may exist in *Streptomyces* sp. We have recently used anti-DtxR antibodies to probe immunoblots of crude extracts of *Brevibacterium lactofermentans*, *S. lividans*, and *Mycobacterium tuberculosis*. In all three instances, we have detected immunoreactive proteins with apparent molecular weights of 26,000-30,000. In the case of *B. lactofermentans,* Oguiza et al[41] have reported the molecular cloning and characterization of a *dtxR* homologue. Indeed, DtxR from *B. lactofermentans* was shown to be an iron-dependent repressor and to control the expression of β-galactosidase from a diphtheria *tox* promoter / operator *lacZ* transcriptional fusion.

In several species of Gram-negative organisms, Fur and Fur-like proteins have been found (e.g., *E. coli, V. cholerae, Y. pestis,* and *P. aeruginosa*),[42-45] and the amino acid sequence of the Fur family of regulatory proteins and the nucleic acid sequence of many of their respective DNA binding sites have been found to be homologous. Accordingly, it is tempting to speculate that DtxR from *C. diphtheriae* is one of a family of transition metal ion-activated regulatory proteins that is present in *Corynebacterium,* as well as the genera *Streptomyces* and *Mycobacterium.* We anticipate that members of the DtxR-family of regulatory proteins will have homologous amino acid sequences, DNA binding targets and they will function in the regulation of iron responsive genes in their respective genera.

ACKNOWLEDGMENTS

This work is partially supported by Public Health Service grant AI-21628 (J.R.M.) from the National Institute of Allergy and Infectious Diseases and by a fellowship from the Wissenschaftausschussder NATO ubër den DAAD (N.S.). We thank Drs. Johanna vanderSpek and Harold Amos for their critical reading of the manuscript.

REFERENCES

1. Litwin CM, Calderwood SB. Role of iron in regulation of virulence genes. Clin Microbiol Rev 1993; 6:137-149.
2. Boyd J, Oza MN, Murphy JR. Molecular cloning and DNA sequence analysis of a diphtheria *tox* iron-dependent regulatory element (*dtxR*) from *Corynebacterium diphtheriae.* Proc Nat Acad Sci USA 1990; 87:5968-5972.
3. Schmitt MP, Holmes RK. Iron-dependent regulation of diphtheria toxin and siderophore expression by the cloned *Corynebacterium diphtheriae* repressor *dtxR* in *C. diphtheriae* C7 strains. Infect Immun 1991; 59:1899-1904.
4. Tao X, Murphy JR. Binding of the metallo-regulatory protein DtxR to the diphtheria *tox* operator reuqires heavy metal ions and protects the palindromic sequence from DNase I digestion. J Biol Chem 1992; 267:21761-21764.
5. Schmitt MP, Holmes RK. Analysis of diphtheria toxin repressor-operator interactions and characterization of a mutant repressor with decreased binding activity for divalent metals. Mol Microbiol 1993; 9:173-181.
6. Neilands JB. Microbial iron compounds. Ann Rev Biochem 1981; 50:715-731.
7. Pappenheimer AM Jr, Johnson SJ. Studies on diphtheria toxin production: I. The effect of iron and copper. Brit J Exp Path 1936; 17:335-341.
8. Pappenheimer AM Jr. Symp Soc Gen Microbiol 1955:40-56.
9. Uchida T, Gill DM, Pappenheimer AM Jr. Mutation in the structural gene for diphtheria toxin carried by temperate phage β. Nature (New Biol) 1971; 233:8-11.
10. Matsuda M, Barksdale L. System for the investigation of bacteriophage-directed synthesis of diphtheria toxin. J Bacteriol

1967; 93:722-730.

11. Gill DM, Uchida T, Singer R. Expression of diphtheria toxin genes by integrated and non-integrated phage beta. Virology 1972; 50:664-668.

12. Matsuda M, Barksdale L. Phage-directed synthesis of diphtheria toxin in non-toxinogenic *Corynebacterium diphtheriae*. Nature 1966; 210:911-913.

13. Uchida T, Pappenheimer AM Jr, Greany R. Diphtheria toxin and related proteins: I. Isolation and properties of mutant proteins serologically related to diphtheria toxin. J Biol Chem 1973; 248:3838-3844.

14. Murphy JR, Skiver J, McBride G. Isolation and partial characterization of a corynebacteriophage β, *tox*-operator constitutive-like mutant lysogen of *Corynebacterium diphtheriae*. J Virol 1976; 18:235-244.

15. Murphy JR, Pappenheimer AM Jr, Tayart de Borms S. Synthesis of diphtheria *tox*-gene products in *Escherichia coli* extracts. Proc Natl Acad Sci USA 1974; 71:11-15.

16. Kanei C, Uchida T, Yoneda M. Isolation from *Corynebacterium diphtheriae* C7(β) of bacterial mutants that produce diphtheria toxin in medium with excess iron. Infect Immun 1977; 18:203-209.

17. Murphy JR, Bacha P. (1979) In: Microbiology 1979. Washington, DC: American Society for Microbiology, 1979:181-186.

18. Welkos SL, Holmes RK. Regulation of toxinogenesis in *Corynebacterium diphtheriae*. I. Mutations in bacteriophage β that alter the effects of iron on toxin production. J Virol 1981; 37:936-945.

19. Welkos SL, Holmes RK. Regulation of toxinogenesis in *Corynebacterium diphtheriae*. II. Genetic mapping of a *tox* regulatory mutation in bacteriophage J Virol 1981; 37:946-954.

20. Krafft AE, Tai SP, Coker C et al. Transcription anaylsis and nucleotide sequence of *tox* promoter/operator mutants of corynebacteriophage β. Micro Path 1992; 13:85-92.

21. Leong D, Murphy JR. Characterization of the diphtheria *tox* transcript in *Corynebacterium diphtheriae* and *Escherichia coli*. J Bacteriol 1985; 163:1114-1119.

22. Boyd J, Murphy JR. Analysis of the diphtheria *tox* promoter by site-directed mutagenesis. J Bacteriol 1988; 170:5949-5952.

23. Kaczorek M, Delpeyroux F, Chenciner N et al. Nucleotide sequence and expression of the diphtheria *tox228* gene in *Escherichia coli*. Science 1983; 221:855-858.

24. Greenfield L, Bjorn MJ, Horn G et al. Nucleotide sequence of the structural gene for diphtheria toxin carried by corynephage β. Proc Natl Acad Sci USA 1983; 80:6853-6857.

25. Ratti G, Rappuoli R, Giannini G. The complete nucleotide sequence of the gene coding for diphtheria toxin in the corynephage ω(tox+) genome. Nucl Acids Res 1983; 11:6589-6595.

26. Fourel G, Phalipon A, Kaczorek M. Evidence for direct regulation of diphtheria toxin gene transcription by an Fe^{2+}-dependent DNA-binding repressor, DtoxR, in *Corynebacterium diphtheriae*. Infect Immun 1989; 57:3221-3225.

27. Tai SP, Krafft AE, Nootheti P et al. Coordinate regulation of siderophore and diphtheria toxin production by iron in *Corynebacterium diphtheriae*. Microb Pathog 1990 9:267-273

28. Boyd JM, Hall KM, Murphy JR. DNA sequence and characterization of *dtxR* alleles from *Corynebacterium diphtheriae* PW8(−), 1030(−), and C7hm723(−). J Bacteriol 1992; 174:1268-1272.

29. Schmitt MP, Holmes RK. Characterization of a defective toxin repressor (*dtxR*) allele and analysis of *dtxR* transcription in wild type and mutant strains of *Corynebacterium diphtheriae*. Infect Immun 1991; 59:3903-3908.

30. Wang Z, Schmitt MP, Holmes RK. Characterization of mutations that inactivate the diphtheria toxin repressor gene (*dtxR*). Infect Immun 1994; 62:1600-1608.

31. Schiering N, Tao X, Murphy JR et al. Crystallization and preliminary studies of the diphtheria *tox* repressor from *Coryne-*

bacterium diphtheriae. J Mol Biol 1994; (in press).

32. Tao X, Boyd J, Murphy JR. Specific binding of the diphtheria *tox* regulatory element DtxR to the *tox* operator requires divalent heavy metal ions and a 9-base pair interrupted palindromic sequence. Proc Natl Acad Sci USA 1992; 89:5897-5901.

33. Groman NB, Judge K. Effect of metal ions on diphtheria toxin production. J Bacteriol 1979; 135:511-516.

34. Tao X, Murphy JR. Cysteine-102 is positioned in the metal binding activation site of the *Corynebacterium diphtheriae* regulatory element DtxR. Proc Natl Acad Sci USA 1993; 90:8524-8528.

35. Creighton TE. In: Proteins: Structure and Molecular Properties. New York: Freeman, 1993:361-363.

36. Schmitt MP, Holmes RK. Cloning, sequencing, and footprint analysis of two promoter / operators from *Corynebacterium diphtheriae* that are regulated by the diphtheria toxin repressor (DtxR) and iron. J Bacteriol 1994; 176:1141-1149.

37. Tao X, Murphy JR. Determination of the DtxR consensus binding site by in vitro affinity selection. Proc Natl Acad Sci USA 1994; 91:9646-9650.

38. Wright WE, Binder M, Funk W. Cyclic amplification and selection of targets (CASTing) for the myogenin consensus binding site. Mol Cell Biol 1991; 11:4104-4110.

39. Günter K, Toupel C, Schupp T. Characterization of an iron-regulated promoter involved in deferrioxamine B synthesis in *Streptomyces pilosus*: repressor-binding site and homology to the diphtheria toxin gene promoter. J Bacteriol 1993; 175:3295-3302.

40. Schupp T, Toupet C, Divers M. Cloning and expression of two genes of *Streptomyces pilosus* involved in the biosynthesis of siderophore desferroxamine B. Gene 1988; 64:179-188.

41. Oguiza JA, Tao X, Marcos T et al. Molecular cloning, DNA sequence analysis and characterization of the *Corynebacterium diphtheriae dtxR* homolog from *Brevibacterium lactofermentum.* J Bacteriol 1994; (in press).

42. Schaffer S, Hantke K, Braun V. Nucleotide sequence of the iron regulatory gene *fur.* Mol Gen Genet 1985; 200:111-113.

43. Litwin CM, Boyko SA, Calderwood SB. Cloning, sequencing, and transcriptional regulation of the *Vibrio cholerae fur* gene. J Bacteriol 1992; 174:1897-1903.

44. Staggs TM, Perry RD. Identification and cloning of a *fur* regulatory gene in *Yersinia pestis.* J Bacteriol 1991; 173:417-425.

45. Prince RW, Cox CD, Vasil ML. Coordinate regulation of siderophore and exotoxin A production: molecular cloning and sequencing of the *Pseudomonas aeruginosa fur* gene. J Bacteriol 1993; 175:2589-2598.

BVGAS DEPENDENT PHENOTYPIC MODULATION OF *BORDETELLA* SPECIES

Peggy A. Cotter, Brian J. Akerley and Jeff F. Miller

1. *BORDETELLA* SPECIES

B*ordetella* species cause respiratory tract infections in humans and other animals. *Bordetella pertussis* and *Bordetella parapertussis* have adapted exclusively to the human host causing whooping cough (pertussis) and a milder pertussis-like disease, respectively. Despite the availability of a vaccine, pertussis continues to be a significant cause of morbidity and mortality in young children throughout the world and was reported to cause 400,000 deaths in 1992.[1] Pertussis is one of the most infectious diseases known; attack rates of 70% to 100% in susceptible household contacts have been reported.[2-4] The disease progresses in three stages. A 10-14 day incubation period is followed by the catarrhal stage characterized by mild, nondistinctive, cold-like symptoms including rhinorrhea, lacrimation, malaise and low-grade fever. After 7-10 days the illness progresses to the paroxysmal phase. Paroxysmal phase symptoms include violent coughing spasms (paroxysms) followed by an inspiratory gasp resulting in the hallmark whooping sound for which the disease is named. This stage lasts from 1 to 4 weeks. Symptoms gradually become less severe and paroxysms less frequent during the convalescent period which may last up to 6 months. *B. pertussis* is believed to spread exclusively via respiratory droplets. Recovery of *B. pertussis* from the nasopharynx of infected individuals is maximal during the catarrhal stage, and it is assumed that transmission potential is greatest at this time, decreasing as the disease progresses.[2] The host antibody response is thought to play a critical role in clearance of *B. pertussis* and protection from reinfection.[5]

Signal Transduction and Bacterial Virulence, edited by Rino Rappuoli,
Vincenzo Scarlato and Beatrice Aricò. © 1995 R.G. Landes Company.

Bordetella bronchiseptica, which is also highly infectious, can infect a variety of mammals. Although asymptomatic colonization of the respiratory tract is the most common result of bordetellosis, *B. bronchiseptica* can cause atrophic rhinitis in pigs, kennel cough in dogs, snuffles in rabbits and bronchopneumonia in guinea pigs.[6] *B. bronchiseptica* can establish persistent infections despite the production of high titers of *Bordetella* specific serum antibodies.[6,7] *B. bronchiseptica* is also believed to be transmitted via respiratory droplets, however, spread by direct contact or via fomites has been suggested since this organism is much less fastidious than *B. pertussis* and can survive and multiply in nutrient poor media and even phosphate buffered saline (PBS).[7-9]

Bordetella avium, the most distantly related of the four species in the *Bordetella* genus, causes highly contagious upper respiratory tract infections in birds known as coryza. Coryza in turkeys and chickens is characterized by oculonasal discharge, mucus accumulation in the trachea, coughing, sneezing, and weight loss.[10-13] Evidence suggests that transmission may occur via direct contact or through contaminated litter and water, but not between cages, arguing against transmission by aerosol droplets.[14-16]

While there are obvious similarities between the infections caused by *Bordetella* species, bordetellosis in animals differs from pertussis in humans in several respects. Pertussis presents as an acute systemic disease followed by clearance of the organism within 4 to 5 weeks and recovery results in long-lived immunity.[17,18] Bordetellosis is usually a chronic, long term infection without systemic effects and the organisms are rarely cleared from the infected host.[6,13] Some of these differences may be attributable to the action of pertussis toxin, which is produced only by *B. pertussis*, or to the presence or absence of other virulence factors. Differences in symptomatology may also contribute to the modes of transmission postulated for the different species.

1.2. PATHOGENESIS

While the complex series of events that occurs during the *Bordetella*-host interaction is not completely understood, several factors which contribute to pathogenesis have been identified. Infection of a susceptible host by *B. pertussis* begins with the colonization phase in which the organisms localize and multiply among the cilia of respiratory epithelial cells. Factors implicated in adhesion include filamentous hemagglutinin (FHA), a large rod-shaped surface protein also found in culture supernatants;[19-21] pertactin (Prn), a 69 kDa outer membrane protein;[22,23] and fimbriae.[24] Symptoms increase in severity as multiplying bacteria synthesize toxins which cause both local and systemic effects. Tracheal cytotoxin (TCT), a muramyl peptide released from growing cells, causes damage to ciliated respiratory epithelial cells leading to cell death and extrusion from the mucosal surface.[25] TCT is thought to be involved in cough production and a role in transmission has been suggested.[26] The role of dermonecrotic toxin (DNT) in pathogenesis is unclear but it may contribute to local tissue damage. Purified DNT causes vascular constriction resulting in ischemic necrosis in guinea pig lungs.[27] Pertussis toxin (PTX) and adenylate cyclase toxin/hemolysin (AC-/HLY) penetrate eukaryotic cells and cause an increase in intracellular cAMP. In vitro these toxins can inhibit chemotactic and phagocytic abilities of polymorphonuclear leukocytes and macrophages.[28-32] Leukocytosis and hypoglycemia have also been attributed to the action of pertussis toxin.[33] Although *Bordetella* have long been considered to be noninvasive pathogens, the ability to enter and survive inside eukaryotic cells has been reported.[34-37]

B. parapertussis and *B. bronchiseptica* synthesize an almost identical set of virulence factors as *B. pertussis*, with the exception of pertussis toxin.[38-40] In contrast, *B. avium* has been reported to produce only DNT,[41,42] TCT,[42] fimbriae[43,44] and a hemagglutinin which has been suggested to play a role similar to that of FHA.[45]

1.3. THE BVGAS REGULON— AN OVERVIEW

As illustrated in accompanying chapters, coordinate regulation of virulence gene expression is an important strategy used by pathogenic bacteria to adapt to diverse environments.[46] Expression of all of the known virulence factors synthesized by *B. pertussis*, except TCT, is coordinately regulated in response to environmental signals (Fig. 2.1). The reversible loss of virulence determinants by *B. pertussis* under certain growth conditions, a phenomenon known as phenotypic modulation, was recognized early this century[47] and was rigorously examined by Lacey in 1960.[48] Using changes in colony morphology, hemolysin production and antigenicity as indicators, Lacey showed that high temperature, and certain ions such as sodium, potassium, halides, formate and nitrate favor the virulent phase while low temperature and ions such as Mg^{+2}, SO_4^{-2}, and mono- and dicarboxylic acids favor avirulent phase growth. Further investigation showed that chlorate anions and nicotinic acid derivatives also resulted in down regulation of virulence factors.[49] The relationship between these signals and the environments encountered by *B. pertussis* is unknown, but evidence accumulated over the last decade suggests that this coordinate regulatory strategy has significant biological relevance.

Phenotypic modulation is mediated by the products of a single genetic locus, *bvgAS* (alternatively designated *vir*). This locus was first identified by Weiss and Falkow as the site of a Tn5 insertion in *B. pertussis* which had simultaneously lost the ability to synthesize PTX, AC/HLY and FHA.[50] Subsequent genetic and biochemical analyses showed that the *bvgAS* products, BvgA and BvgS, are members of the "two-component" family of signal transduction proteins.[51-53] Such systems allow bacteria to efficiently sense and adapt to changes in their environment.[54,55] In the laboratory, BvgAS is active when cells are grown on artificial medium such as Bordet-Gengou-blood agar, and induces the formation of small, domed, hemolytic

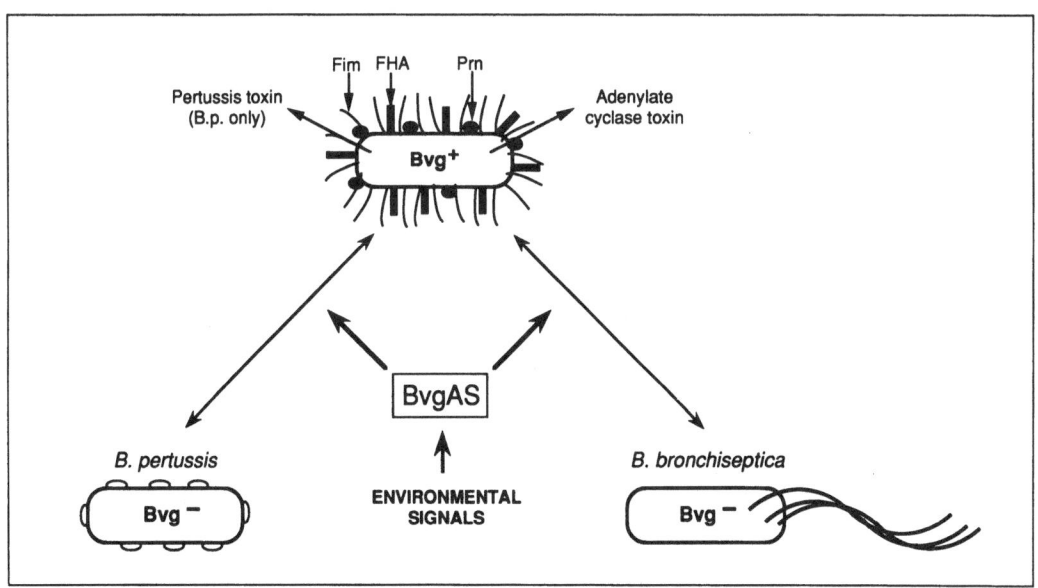

Fig. 2.1. Phenotypic modulation in Bordetella species. Bordetella alternate between the Bvg⁺ and Bvg⁻ phases in response to environmental signals by a process mediated by the BvgAS signal transduction system. In the Bvg⁺ phase adhesins, including fimbriae (Fim), filamentous hemagglutinin (FHA) and pertactin (Prn) and toxins, including pertussis toxin and adenylate cyclase toxin are expressed. Bvg⁻ phase characteristics include the expression of specific outer membrane proteins ( ) in B. pertussis and the synthesis of flagella and the phenotype of motility in B. bronchiseptica.

colonies that express all of the known virulence factors. In response to modulating conditions such as low temperature (25°C vs. 37°C) or the presence of $MgSO_4$ or nicotinic acid, BvgAS is inactive and the organisms switch to the Bvg⁻ phase, stop synthesizing adhesins and toxins, and form large, flat, nonhemolytic colonies. Spontaneous avirulent mutants that arise from virulent phase cultures, called phase variants, result from genetic alterations at the *bvgAS* locus.[52]

While investigating the phenomenon of phenotypic modulation, Lacey recognized that a few antigens were present only under modulating conditions[48] and more recently Beattie et al showed that a subset of outer membrane proteins synthesized by *B. pertussis* are specific to the Bvg⁻ phase.[56] Thus, Bvg⁻ phase cells do not merely lack virulence factors, but acquire proteins that are unique to this phase. BvgAS, therefore, does not function solely to turn off virulence factors but controls the expression of Bvg⁻ phase specific factors as well. Tn*phoA* mutagenesis was used to identify genes (*vrgs*, or *vir* repressed genes) in *B. pertussis* whose expression is increased under modulating conditions.[57] Several *vrg* loci were identified, as well as some new *vir* activated genes (*vags*), and their functions are currently being investigated.

Interestingly, Bvg⁻ phase specific proteins were not readily detected in outer membrane preparations of *B. bronchiseptica* (Cotter PA and Miller JF, unpublished data) and, of the *vrgs* from *B. pertussis* tested so far, none appear to be expressed in *B. bronchiseptica* even though hybridizing sequences are present in the chromosome.[58] In contrast, *B. bronchiseptica* and *B. avium* are motile in the Bvg⁻ phase; but while *B. pertussis* and *B. parapertussis* contain sequences which hybridize to the *B. bronchiseptica* flagellin gene, *flaA*, these pseudogenes are transcriptionally silent and these species are nonmotile.[59] In addition, *B. bronchiseptica* synthesizes an hydroxamate siderophore, alcaligin, which is produced only in the Bvg⁻ phase.[60] Thus, although the Bvg⁺ phase factors synthesized

by *B. pertussis*, *B. parapertussis* and *B. bronchiseptica* are very similar, these species can be divided into two groups based on expression of Bvg⁻ phase phenotypes (Fig. 2.1). To date, no Bvg⁻ phase specific factor common to all *Bordetella* species has been identified.

The ability of *Bordetellae* to alternate between distinct phenotypic phases has been known for many years and considerable progress has been made toward understanding how this switch occurs at the molecular level. Yet, the selective advantage provided by this ability and the role of BvgAS mediated signal transduction in pathogenesis continues to be a major puzzle. While it is clear that Bvg⁺ phase specific factors are required for virulence, the function of the Bvg⁻ phase has not yet been identified. Postulated roles for modulation to the Bvg⁻ phase include establishment of infection, modulation of disease, persistence in the host due to down regulation of antigens, survival in microenvironments within the host and transmission to new hosts. Speculation as to the role of BvgAS is additionally hampered by the fact that, despite rigorous examination of the signals to which BvgAS responds in the laboratory, the true signals recognized by BvgAS in nature are unknown. In this review we present a comparative analysis of BvgAS regulation in *Bordetella* species, with particular emphasis on *B. pertussis* and *B. bronchiseptica*, in an attempt to understand the role of BvgAS mediated signal transduction in *Bordetella* pathogenesis.

2. MECHANISTIC INTERACTIONS BETWEEN BvgA AND BvgS

Considerable effort has focused on understanding how BvgA and BvgS function at the molecular level. This information is important for two reasons. First, understanding the peculiarities of this particular signal transduction system, such as whether it functions as an "on/off" switch or is able to effect a range of phenotypes, could yield insight into the role of BvgAS in pathogenesis. Secondly, understanding how BvgA and BvgS function mechanistically may allow us to generate tools useful for addressing specific questions about the role of BvgAS. For example, Bvg constitutive mutations, the isolated and characterized in *B. pertussis*, led to the construction of "phase locked" *B. bronchiseptica* strains which were used in animal models to test the importance of signal transduction in vivo.[7]

2.1. GENETIC ANALYSIS OF *BVGAS*

Transposon insertion mutations as well as deletions and point mutations in *bvgAS* result in the Bvg⁻ phenotype, demonstrating the requirement of BvgAS for virulence factor expression.[50,52] To assess whether *bvgAS* controls gene expression at the transcriptional level, a fusion between the FHA structural gene (*fhaB*) and the *E. coli* β-galactosidase gene (*lacZ*) contained on a specialized lambda phage was used. When supplied in trans on a multicopy plasmid, *bvgAS* activated expression of the *fhaB'-lacZ* fusion in *E. coli* in a signal dependent manner, showing that BvgAS is sufficient for induction of virulence factor expression in *E. coli*.[61] Proof that BvgAS mediates signal transduction in response to environmental conditions, however, came from the isolation and characterization of constitutive mutations which are unresponsive to signals in *Bordetella* and in *E. coli*.[62] These mutations have been useful in studying various aspects of BvgAS regulation and will be discussed in later sections.

As mentioned above, sequence analysis revealed that BvgA and BvgS belong to the "two-component" signal transduction system family.[52,63] A common theme among these systems is communication of information via a phosphorylation cascade involving conserved "transmitter" and "receiver" domains present on sensor and response regulator proteins, respectively. According to the current paradigm, the membrane bound sensor protein autophosphorylates at a conserved histidine in the transmitter domain in response to environmental signals. The phosphorylated

transmitter domain then serves as a substrate for phosphorylation of a conserved aspartic acid on the receiver domain of the response regulator resulting in activation of this protein (for a review see Parkinson and Kofoid).[54] BvgS is a 135 kDa protein which is located in the cytoplasmic membrane.[53] Similar to other sensor proteins, BvgS contains an N-terminal periplasmic domain, postulated to be involved in signal recognition, which is held in place by flanking transmembrane segments. The second transmembrane segment is followed by a linker region and a transmitter domain which contains the predicted site of autophosphorylation. BvgS is a member of a subfamily of sensor proteins that includes FrzE of *Myxococcus xanthus*, ArcB of *Escherichia coli* and VirA of *Agrobacterium tumefaciens*. These proteins all contain additional domains at their C-termini.[64-66] The transmitter domain of BvgS is followed by an alanine/proline rich segment, a receiver domain, another alanine/proline rich segment, then a C-terminal domain. The alanine/proline rich segments, similar to those found in FrzE of *M. xanthus*, are thought to function as conformationally flexible hinges allowing intramolecular interaction between domains.[66] BvgA is a 23 kDa protein typical of response regulators with an N-terminal receiver domain, containing the putative site of phosphorylation, and a C-terminal helix-turn-helix motif, which is thought to recognize and bind specific DNA sequences.

To investigate the roles of the specific domains of the BvgA and BvgS proteins, targeted mutations have been constructed and analyzed. In-frame linker insertion mutations in the periplasmic domain of BvgS abolish BvgAS activity in vivo.[62,67] Mutations in the transmitter domain of BvgS also abolish function as would be expected since this domain contains the putative site of autophosphorylation and is essential for signal transduction in other systems.[67,68] The receiver domains at the C-termini of ArcB of *E. coli* and VirA of *A. tumefaciens* are postulated to play autoinhibitory roles.[64,69] Deletions or small insertion mutations in the BvgS receiver and C-terminus as well as precise deletions of the BvgS receiver or C-terminus abolish, rather than increase, BvgAS activity indicating that these domains are required for function and are not simply autoinhibitory (Beier and Gross, unpublished data).[62,67,68] Genetic analysis therefore demonstrates a requirement for the BvgS periplasmic, transmitter, receiver and C-terminal domains for BvgAS mediated transcription activation.

Although environmental sensing is thought to be mediated by the periplasmic domain, mutations mapping to this domain which specifically alter the ability

of BvgS to respond to signals have not been described. However, several Bvg-constitutive (BvgS[c]) mutations, which render the system insensitive to environmental signals have been isolated and result from single base pair substitutions in the linker region (Fig. 2.2).[62] Similar mutations have been described in the *E. coli* nitrate sensor, NarX,[70,71] and the chemoreceptor, Tsr.[72] BvgS[c] mutations are dominant to in-frame deletion mutations in the periplasmic domain[62] but not to mutations in the transmitter, receiver or C-terminal domains of BvgS (Uhl MA and Miller JF, unpublished data), demonstrating the importance of the linker in transduction of signals from the environment to the cytoplasm.

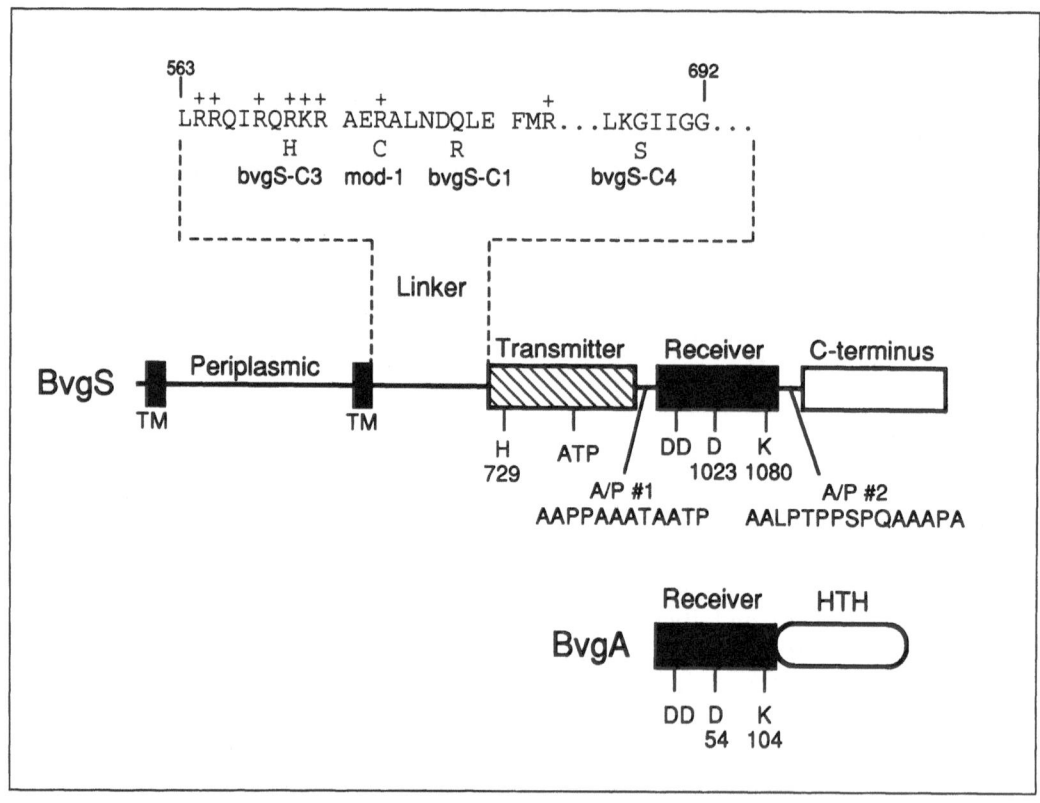

Fig. 2.2. Schematic of the structure of BvgA and BvgS.[51,53] Regions sharing sequence similarity with two component systems are identified. Abbreviations are as follows: TM, transmembrane segments; ATP, consensus ATP binding motif; A/P, alanine/proline rich sequences; HTH, helix turn-helix motif. Conserved transmitter and receiver domains and the C-terminal domain of BvgS are labeled. The sizes of the domains of BvgS are as follows: periplasmic region, 514 aa; linker, 159 aa; transmitter, 232 aa; receiver, 131 aa; C-terminus, 126 aa. Conserved amino acids are shown for BvgA and BvgS. A partial amino acid sequence of the BvgS linker region is shown and the locations of amino acid substitutions conferring insensitivity to environmental signals (Bvg[c] phenotype) are indicated.[62]

Normally, both *bvgA* and *bvgS* are required for activation of virulence gene expression in *B. pertussis* and *E. coli*.[53,61,73] Overproduction of BvgA in *E. coli*, however, results in BvgS-independent activation of *fhaB-lacZ* expression that is unresponsive to environmental signals.[73] Thus, the requirement for activation of BvgA by BvgS is eliminated if BvgA is present in high concentration. Mutations separating the functions of BvgA localize DNA binding ability to the C-terminal domain, which contains the helix-turn-helix motif.[74] The N-terminal receiver domain is required for BvgS mediated activation, which is predicted to occur by phosphorylation of a conserved aspartate residue.[74] Chemical stability analysis of phosphorylated BvgA is consistent with phosphorylation at an aspartic acid[68,74] but specific mutations at the putative site of phosphorylation have not been described. BvgA constitutive mutations that can activate an *fhaB-lacZ* fusion in *E. coli* independent of BvgS have recently been isolated and map near the conserved aspartic acid residue (Uhl MA and Miller JF, unpublished data). These mutations presumably change the conformation of the protein in a way that mimics a phosphorylated state and is competent to activate transcription by acquiring increased affinity for DNA and/or increased ability to interact with RNA polymerase.

The formation of dimers or higher order multimers of BvgA and/or BvgS may be integral to their function. Multimerization has been demonstrated for other two-component systems.[75-78] Mutations in the periplasmic region or the transmitter of BvgS can be complemented by *bvgS* alleles with mutations in the C-terminus suggesting that multimerization of BvgS occurs.[67] Evidence consistent with the formation of BvgA dimers has also been reported.[79]

2.2. BIOCHEMICAL ANALYSIS OF BVGAS

The ability of BvgS and BvgA to participate in a phosphorelay cascade was investigated in an in vitro phosphorylation assay system using BvgA and a derivative of BvgS ('BvgS) missing the transmembrane and periplasmic domains.[68] By analogy with other systems, BvgS is predicted to autophosphorylate at the conserved histidine residue in its transmitter domain (His729), and possibly at the conserved aspartic acid residue in its receiver domain (Asp 1023). The postulated site of phosphorylation of BvgA is the aspartic acid residue at position 54. 'BvgS, was shown to be capable of autophosphorylation with the γ phosphate of ATP, and transfer of phosphate to wild type BvgA (Fig. 2.3).[68] Autophosphorylation, as well as subsequent phosphoryl group transfer to BvgA, was abolished when the conserved histidine residue of BvgS (His 729) was replaced with glutamine, demonstrating the requirement of this residue for phosphorylation and phosphorelay. In contrast, replacement of the aspartate residue in the receiver of 'BvgS (Asp 1023) with asparagine resulted in a protein that was capable of autophosphorylation but was unable to transfer phosphate to BvgA, thereby uncoupling BvgS autophosphorylation from phosphotransfer.[68] The lysine at position 1080 is also conserved among receiver domains and changes at this residue (Lys1080-Arg and Lys1080-Met) also resulted in lack of transfer to BvgA, with retention of autophosphorylation activity that is indistinguishable from that of wild type 'BvgS (Uhl MA and Miller JF, unpublished data). These results indicate that a phosphorylation competent BvgS receiver domain is required for phosphotransfer.

While full understanding of the mechanisms by which environmental signals are converted to virulence gene regulation has not yet been achieved, the basic tenets of BvgAS signal transduction have been illustrated and a model has been proposed (Fig. 2.3).[68] Accordingly, BvgS resides in the cytoplasmic membrane and under modulating (Bvg⁻ phase) conditions, remains unphosphorylated and does not activate gene transcription. The low level of BvgA, which is mostly unphosphorylated,

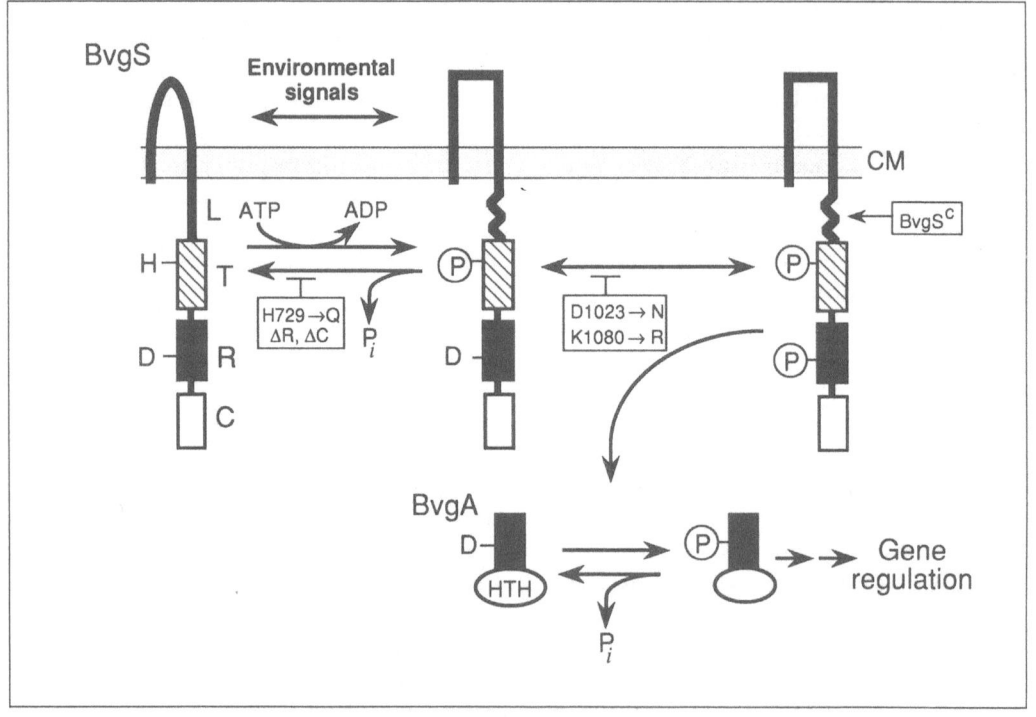

Fig. 2.3. Model, adapted from Uhl and Miller,[68] showing proposed steps in BvgAS signal transduction. Environmental signals (temperature, $MgSO_4$, and nicotinic acid) regulate the activity of BvgS. In its active form, BvgS autophosphorylates at His-729. Following intramolecular phosphoryl group transfer to Asp-1023, BvgS transfers a phosphate to BvgA at the conserved Asp residue. Phosphorylated BvgA is then capable of activating or repressing gene transcription. Mutations that alter the site of phosphorylation (H729→Q) or remove the receiver or C-terminus of BvgS (DR, DC) prevent autophosphorylation. Mutations at the conserved aspartate (D1023→N) and lysine (K1080→R) of the receiver domain of BvgS are necessary for phosphotransfer to BvgA. Signal insensitive mutations (BvgS^c) map to the linker. Abbreviations: L, linker; T, transmitter; R, receiver; C, C-terminal domain; CM, cytoplasmic membrane. Reprinted with permission from Uhl MA and Miller, Proc Natl Acad Sci USA 1994; 91:1163-1167.

is unable to induce expression of virulence genes (*vags*) or to mediate repression of Bvg repressed genes (*vrgs*). In response to Bvg+ phase conditions, such as a shift to 37°C, BvgS autophosphorylates at His729 of the transmitter domain. This phosphate is then intramolecularly transferred to the Asp1023 residue of the receiver domain of BvgS. Phosphorylation of the receiver of BvgS appears to be required for subsequent phosphotransfer to BvgA. Phosphorylation of BvgA at the Asp54 residue of its receiver renders it competent to activate transcription of virulence genes and cause repression of Bvg repressed loci. This working model is intended as a framework for developing and testing new hypotheses regarding details of BvgAS sig-

nal transduction. The roles of the "extra" domains on BvgS remain to be determined. It is tempting to speculate that they may be involved in allowing the system to differentially regulate target gene expression to reflect the integration of a number of signals and/or signals of different magnitudes.

2.3. COMPARATIVE ANALYSIS OF BVGAS

The predicted amino acid sequences of the BvgA and BvgS proteins of *B. pertussis*, *B. parapertussis* and *B. bronchiseptica* are highly conserved, with 96% amino acid identity.[51] BvgA proteins of all three species have the same amino acid sequence. Interestingly, the BvgS proteins differ primarily in the periplasmic domain, where 68% of the total amino acid changes occur, yet these three species respond to the same set of environmental signals. Of 198 bp changes observed in the *bvgAS* locus, 173 (87%) are identical in *B. parapertussis* and *B. bronchiseptica,* consistent with the proposed phylogenetic assignment of *B. bronchiseptica* and *B. parapertussis* being more closely related to each other than to *B. pertussis*.[51,80-82] Sequences which hybridize to *bvgS*, but not to *bvgA*, have been demonstrated in *B. avium*,[83] but it's assumed that *B. avium* has a similar locus since it synthesizes factors which are BvgAS regulated in other species (DNT and motility) and these factors are regulated by environmental signals in a similar manner.[42,84]

Biochemical analysis has been performed only with BvgA and BvgS from *B. pertussis*. Genetic studies show that deletions and disruptions of *bvgS* in *B. bronchiseptica* abolish function as in *B. pertussis*[51,85] Additionally, the BvgS-C3 mutation isolated and characterized in *B. pertussis*[62] was introduced into *B. bronchiseptica* and it conferred the same BvgS-constitutive phenotype in this species as in *B. pertussis*.[7,84] While the BvgAS loci of these species appear to be almost identical and functionally interchangeable, complementation studies reveal some interesting differences. Multiple copies of *bvgAS* from either *B. pertussis* or *B. bronchiseptica* can complement *bvgS* mutations in *B. pertussis* but appear to only partially complement *bvgS* mutations in *B. bronchiseptica* (Cotter PA and Miller JF, unpublished data).[51,85] Multiple copies of *bvgS* in *B. bronchiseptica* resulted in hyperproduction of BvgA and it was suggested that this leads to inactivation of BvgAS.[51] Multiple copies of *bvgS* did not result in hyperproduction of BvgA in *B. pertussis*. Since the *bvgAS* clones from either *B. pertussis* or *B. bronchiseptica* functioned differently in the two species these differences may be due to factors other than *bvgAS* which influence its regulatory functions.

3. TRANSCRIPTIONAL REGULATION OF BvGAS CONTROLLED GENES

Sensory transduction can be conceptually divided into three steps; recognition of environmental signals, transduction of this information to the cytoplasm, and induction of either phenotypic or behavioral changes. Investigation of the last step of this process, control of gene expression by BvgA and BvgS, has perhaps yielded the most insight into the role of BvgAS mediated signal transduction in *Bordetella*. Genes activated by BvgAS include those encoding adhesins (*fhaB, prn,* and *fim*) and toxins (*ptx, cya* and *dnt*) as well as those required for production of cytochrome *d*-629, capsule formation, several uncharacterized outer membrane proteins,[86,87] and *bvgAS* itself. Genes whose expression is repressed by BvgAS (*vrg*s) include genes identified by Tn*phoA* mutagenesis in *B. pertussis*,[57] and the *frlAB* and *flaA* loci of *B. bronchiseptica* which are involved in flagella synthesis.[59,84] Current information regarding control of gene expression at each of the relevant promoters will be discussed as it relates to understanding BvgAS signal transduction.

3.1. BvGAS-ACTIVATED GENES

3.1.1. *fhaB* and *bvgAS*

Of the BvgAS activated genes examined, *fhaB* and the *bvgAS* operon are thought to be directly activated by BvgA based on several lines of evidence. Both *fhaB* and *bvgA* are able to respond to BvgA and BvgS in *E. coli*.[61,88] As mentioned earlier, *bvgAS* supplied in trans on a multicopy plasmid activates *fhaB'-lacZ* expression in *E. coli*. *bvgA* expression is also induced by *bvgAS* in *E. coli*, as well as in *B. pertussis*, demonstrating positive autoregulation of this operon.[89] BvgAS mediated expression of *fhaB* and *bvgA* decreases in response to the presence of MgSO$_4$, nicotinic acid or low temperature in *E. coli* just as in *B. pertussis*,[61] and the transcriptional start sites for *fhaB* and *bvgA* are the same in *E. coli* as in *B.*

pertussis.[61,79,89] These results demonstrate reconstitution of BvgAS dependent signal transduction in *E. coli* and indicate that BvgA and BvgS are the only *Bordetella* specific factors required for activation of these genes.

fhaB and *bvgAS* are adjacently situated as divergently transcribed operons in *B. pertussis*. Primer extension analysis and S1 nuclease protection analysis revealed the presence of several transcriptional start sites in the region between *fhaB* and *bvgAS*.[79,89] In the Bvg⁻ phase, induced either by the presence of modulating signals or by mutation of *bvgAS*, transcription of *bvgAS* initiates at the *bvg*P2 promoter, located 143 bp 5' of the structural gene for *bvgA*. This relatively weak promoter presumably maintains a low, basal level of BvgA and BvgS in the cell under modulating conditions. In the Bvg⁺ phase, transcription of *bvgA* begins primarily from *bvg*P1, located 93 bp 5' of the BvgA ATG codon.[79,89] Transcription from *bvg*P3, located 271 bp 5' of the BvgA initiation codon, and *bvg*P4, which produces an antisense transcript, is also induced under Bvg⁺ phase conditions but the contribution of these promoters to regulation of *bvgAS* expression is unclear. The *fhaB* promoter, located 70 bp upstream from the initiation codon for the structural gene, is activated only under Bvg⁺ phase conditions.

Deletion analysis, DNaseI protection, and gel retardation studies identified 7 bp *cis*-acting elements located in the intergenic region between *bvgAS* and *fhaB* which are required for BvgAS dependent activation of these genes.[73] Two of these elements are directly repeated upstream of the *bvg*P1 promoter, between the -10 and -35 sequences of the *bvg*P2 promoter, and two more, in inverted orientation, are located upstream of the *fhaB* promoter.[90] When expressed on a high copy number plasmid, an oligonucleotide encoding the elements in inverted orientation inhibits *bvgAS* activation of an *fhaB'-lacZ* fusion in *E. coli*, presumably due to titration of BvgA.[91] These results support genetic data

implicating BvgA as a transcriptional regulator that binds specific DNA sequences. Additionally, the location of the control elements at the *bvg*P1 and P2 promoters likely accounts for the switch in expression from P2 to P1 observed in response to BvgA activation, and thus BvgA appears to function as both an activator and a repressor of transcription.

3.1.2. *ptx*

The genes encoding pertussis toxin are organized as an operon (*ptx*) which is positively regulated by BvgAS.[92,93] In contrast to *fhaB* and *bvgAS*, expression of a *ptx'-lacZ* fusion is not efficiently activated by *bvgAS* in mid-exponential phase cultures of *E. coli*.[61] Analysis of the promoter region showed that sequences located 170 bp upstream of the initiation codon are required for BvgAS dependent activation of *ptx* expression in *B. pertussis*,[94] but control elements similar to those identified upstream of *fhaB* and *bvgA* were not found within these sequences.[61,95] Additionally, binding of BvgA to sequences upstream of *ptx* could not be demonstrated.[90,95] These data led to the suggestion that additional factors may be required for induction of *ptx* transcription, but searches for such factors have resulted only in mutations mapping to *rpoA* [96] or *bvgAS*.[67,97] The discrepancy between the lack of *ptx-lacZ* expression in *E. coli* and the inability to find additional regulatory factors may have been partially resolved by recent observations that in fact *ptx* expression can be induced in *E. coli* under certain conditions. Scarlato et al reported activation of a plasmid encoded *ptx-cat* fusion using primer extension as an assay for expression.[98] Activation was enhanced by the addition of DNA gyrase inhibitors, but response to modulators was variable. In our laboratory BvgAS-dependent induction of a chromosomally located *ptx-lacZ* fusion in *E. coli* was demonstrated: *ptx-lacZ* expression was readily detected during slower growth rates (Uhl MA and Miller JF, submitted for publication). *ptx-lacZ* expression decreased to background levels in response to the presence of MgSO₄ or nicotinic acid in the growth medium. Taken together, these data may eliminate the need to invoke an absolute requirement for additional *Bordetella*-specific regulatory factors.

3.1.3 *cya*

The *cyaABDE* operon encodes the bifunctional adenylate cyclase/hemolysin toxin as well as proteins required for its secretion.[99] A BvgAS-dependent promoter located 115 bp upstream of the initiation codon for *cyaA* transcribes the *cyaA* gene.[101] The rest of the operon is transcribed from a second, BvgAS-independent promoter, located within the intergenic region between *cyaA* and *cyaB* as well as from the *cyaA* promoter.[101] Sequences required for BvgAS control of the *cyaA* gene are located within 569 bp of the promoter but, as for *ptx*, elements similar to those found upstream of *bvgAS* and *fhaB* are not present.[102] Additionally, expression of a *cyaA'-lacZ* fusion was not activated by BvgAS in mid-log phase cultures of *E. coli*.[102] The *cyaC* gene, encoding an enzyme required for activation of the toxin, is located upstream of *cyaABDE* and is transcribed in the opposite direction.[100] Regulation of cyaC expression has not been reported.

3.1.4 *fim*

B. pertussis is capable of synthesizing fimbriae of two different serotypes. The structural genes encoding the fimbrial subunits, *fim2* and *fim3*, have been cloned and sequenced and are regulated by BvgAS.[103,104] In addition, *B. pertussis* contains a silent fimbrial subunit gene, *fimX*.[44] Genes required for export and assembly of fimbrial subunits, *fimB-D* are located downstream of *fhaB* and appear to be transcribed from the *fhaB* promoter as well as from another BvgAS dependent promoter located upstream of *fimB*.[105] Sequences similar to the control elements located upstream of *fhaB* and *bvgAS* are located approximately 90 bp upstream of the *fimB* promoter.[105] In addition to BvgAS-dependent regulation,

expression of fimbrial genes is subject to phase variation.[105a] Fimbriae have been implicated as being important for colonization of the trachea of *B. pertussis* infected mice.[24]

3.1.5 *prn*

B. pertussis produces a 69 kDa outer membrane protein, pertactin, that is thought to be involved in adhesion to eukaryotic cells.[22] The structural gene, *prn*, has been cloned and sequenced and is positively regulated by *bvgAS*.[87] Although a detailed characterization of the *prn* promoter region has not been reported, extrapolation from other data suggests *prn* may be activated directly by BvgA. The *rpoA* mutation described by Carbonetti et al,[97] identified in a search for intermediate regulatory loci, affects *ptx* and *cyaA* expression but not *fhaB*, *fim* or *prn* expression. This result suggests that *prn* is regulated similarly to *fhaB* and *fim*, both of which are thought to be directly activated by BvgA.

3.2. DIFFERENTIAL CONTROL OF BVGAS ACTIVATED GENES

Based on transcriptional analyses described above, BvgAS activated genes can be divided into two classes; those whose expression appears to be directly activated by BvgA (*bvgAS*, *fhaB*, *fim* and *prn*) and those in which activation of transcription appears to be more complex (*ptx* and *cya*). A study of the kinetics of virulence gene induction in response to BvgAS activation supports this classification.[106] Transcriptional activation of *fhaB*, *ptx*, *cyaA* and *bvgAS* was assessed by S1 nuclease mapping at various time points after a temperature shift from 25°C (modulating conditions) to 37°C (nonmodulating conditions) (Fig. 2.4). *fhaB* transcription was induced almost immediately after the temperature shift. Transcription from the *bvg*P1 promoter was also induced within 10 minutes following the temperature shift and continued to increase for 2 hours, while transcription from the *bvg*P2 promoter decreased. In striking contrast, tran-

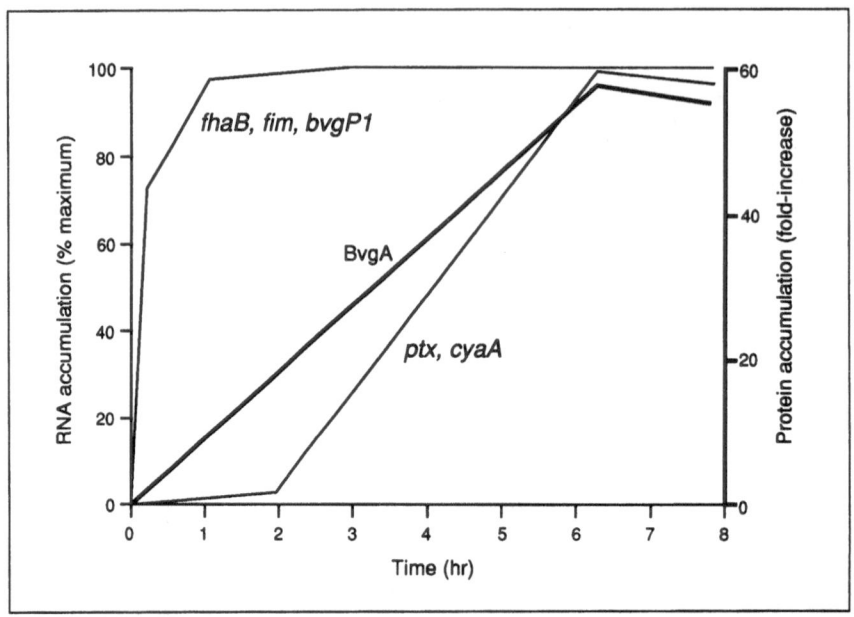

Fig. 2.4. A schematic representing the kinetics of virulence gene induction in response to
BvgAS activation as described by Scarlato et al.[106,107] B. pertussis *cultures were shifted
from 25 °C to 37 °C at time 0, at which point aliquots were taken, mRNA prepared, and
promoter activity assessed by S1 nuclease mapping. The graph represents quantitation of
the results by densitometry. The responses of the fhaB, fim and bvgAP1 promoters were
similar and are represented by one line, as are those of ptx and cyaA. While fhaB, fim and
bvgAP1 expression was induced immediately, reaching maximal levels within 1 hour
postshift, ptx and cyaA expression was not induced until 2 hours after the temperature
shift and did not reach maximal levels until 6 hours postshift. Accumulation of BvgA
protein following a shift from 25 °C to 37 °C, as determined by western blot analysis, is also
shown. Reprinted with permission from Scarlato VB et al. EMBO 1991; 10:3971-3975.*

scription of *ptx* and *cyaA* was not induced
until 2 hours after the temperature shift
and maximal levels were not reached for
another 4 hours. Transcriptional activation of *ptx* and *cyaA* was shown to correlate with increasing BvgA concentration
as determined by Western blot analysis;
BvgA levels increased gradually after the
temperature shift, reaching a maximum
at about 6 hours. Scarlato and cowork-
ers concluded that BvgAS activation of
virulence gene expression occurs in two
steps.[106,107] Genes required for adhesion to
eukaryotic cells, FHA and fimbriae, as well
as *bvgAS* itself, are activated in the first
step. In the second step genes encoding
toxins required for virulence, *ptx*, *cyaA* and
possibly others, are activated. It was proposed that the second step depends on
the intracellular concentration of BvgA,

possibly reflecting a decreased affinity of
BvgA for binding at these promoters.

Thus, BvgA may function differently
at various promoters and may be able to
affect both qualitative as well as quantitative differences in virulence factor production. Rappuoli and colleagues[106,107]
proposed a model for temporal regulation
of virulence factors, suggesting that BvgAS
may become activated as the bacterium
enters a new host and that the upshift in
temperature results in sequential induction of factors required for adhesion and
virulence. Alternatively, or in addition,
BvgAS may be capable of differentially
regulating gene expression in response to
incremental changes in steady state levels
of environmental signals. Melton and
Weiss used *ptl-lacZ* and *fhaB-lacZ* fusions
generated by Tn5*lac* to investigate BvgAS

activity in response to various concentrations of modulating compounds.[108] The *ptl* operon encodes genes required for pertussis toxin secretion and is activated by BvgAS.[109] *ptl-lacZ* expression remained at high, constant levels as the concentration of MgSO$_4$ in the culture medium increased until a modulating dose was reached, then expression dropped to that of the same fusion in a Bvg⁻ strain. In contrast, *fhaB-lacZ* expression decreased steadily in response to increasing concentrations of MgSO$_4$. Thus, the *ptl-lacZ* fusion exhibited an all-or-none response suggesting a threshold level of BvgA activity is required for *ptl* induction, while *fhaB-lacZ* induction varied incrementally with MgSO$_4$ concentration. Similar effects were observed when protein profiles of whole cell extracts of *B. bronchiseptica* grown in the presence of various MgSO$_4$ concentrations were analyzed by Western blot with serum from *B. bronchiseptica* infected rabbits (Cotter PA and Miller JF, unpublished data). While the amounts of some BvgAS regulated proteins appeared to vary in parallel with MgSO$_4$ concentration, others remained at constant levels up to a threshold concentration of MgSO$_4$ then disappeared. These results indicate that BvgAS can differentially activate gene expression in response to the same environmental signals. Taken together, these data demonstrate the complexity of BvgAS signal transduction and suggest that this system may allow *B. pertussis* to sense and respond appropriately to a myriad of environments which may be encountered during pathogenesis by coordinately and precisely regulating gene expression.

4. BvGAS-REPRESSED GENES

4.1. *VRGS* IN *B. PERTUSSIS*

Five *vrg*s identified by Tn*phoA* mutagenesis of *B. pertussis* have been characterized at the molecular level.[56,110] Northern blot analysis indicates that control is at the level of transcription. Deletion and linker insertion analysis identified sequences within the coding regions that are important for BvgAS dependent control of transcription. A 62 bp fragment containing these sequences was used in Southwestern experiments and identified a 34 kDa protein which was less abundant under modulating conditions, suggesting that it may represent an intermediate regulatory protein.[110] It was postulated that this protein is a repressor whose expression is induced by BvgAS under Bvg⁺ phase conditions.

4.2 *VRGS* IN *B. BRONCHISEPTICA*

Considerable progress has been made toward understanding the Bvg⁻ phase of *B. bronchiseptica*, based in part on the investigational accessibility of the predominant phenotype of motility. Two motility loci have been characterized in *B. bronchiseptica*, *flaA* (the structural gene for flagellin) and *frlAB* (a master regulatory locus controlling motility, similar to *flhDC* of *E. coli*).[84] Transcription of both *flaA* and *frlAB* is negatively regulated in the Bvg⁺ phase.[59,84] Based on complementation studies, and by analogy with the motility regulon of *E. coli*, *frlAB* is proposed to function as a transcriptional activator at the top of a regulatory cascade required for synthesis and assembly of flagella as well as proteins required for chemotaxis.[59] Alteration of the *frlAB* promoter such that its expression is activated by BvgA results in ectopic synthesis of flagella and motility in the Bvg⁺ phase, confirming that activation of *frlAB* is sufficient to induce the motility regulatory cascade.[111] Thus, *frlAB* represents the first characterized intermediate regulatory locus of the BvgAS regulon. Transcription of *frlAB* is thought to be directly repressed by BvgA, and sequences with similarity to the putative BvgA binding sites upstream of *fhaB* and *bvgAS* can be found within the region upstream of the *frlAB* promoter. Whereas motility in *E. coli* is ultimately controlled by cAMP and CRP, and thus responds to hunger signals, motility in *B. bronchiseptica* is controlled by *bvgAS* and is therefore linked to pathogenesis. The role of BvgAS regulation of motility in pathogenesis is currently being investigated and will be addressed in a later section. Although *B. pertussis* is nonmotile and contains silent *frlAB* and *flaA* loci, *frlAB* from *B. pertussis* was functional when expressed in *B. bronchiseptica* (Akerley BJ and Miller JF, unpublished data). The role of *frlAB* in *B. pertussis* is unknown.

5. EXPERIMENTAL APPROACHES TO UNDERSTANDING THE ROLE OF BvgAS MEDIATED SIGNAL TRANSDUCTION IN THE *BORDETELLA* LIFE CYCLE

Infection by *B. pertussis* begins when the organism enters a susceptible host and colonizes the ciliated cells of the nasal epithelium. The bacteria remain localized to the epithelial surface as the infection spreads to the lower respiratory tract.

At some point, *B. pertussis* must be transmitted to a new host. The postulated mode of transmission, directly from person to person via respiratory droplets is consistent with the belief that this fastidious organism is unable to survive for long periods of time in the environment. In contrast, *B. bronchiseptica* and *B. avium*, the two *Bordetella* species which are motile in the Bvg⁻ phase, are much heartier organisms. These species may be able to survive longer periods of time outside the host and environmental reservoirs have been suggested for both. *B. pertussis* is thought to be unable to survive for long periods of time in the environment and the postulated mode of transmission, directly from host to host via aerosol droplet, suggests that the time spent between hosts is very short. In contrast, *B. bronchiseptica* and *B. avium*, the two *Bordetella* species which are motile in the Bvg⁻ phase, are much less fastidious than *B. pertussis* and *B. parapertussis*. These organisms may be able to survive longer periods of time outside the host and environmental reservoirs have been suggested for both. Thus, the life cycles of the various *Bordetella* species appear to be similar during the time spent in the host and to differ primarily during the time spent in transit from host to host (Fig. 2.5). It is tempting to speculate that differences in the Bvg⁻ phases of these organisms may reflect differences in the environments occupied between hosts.

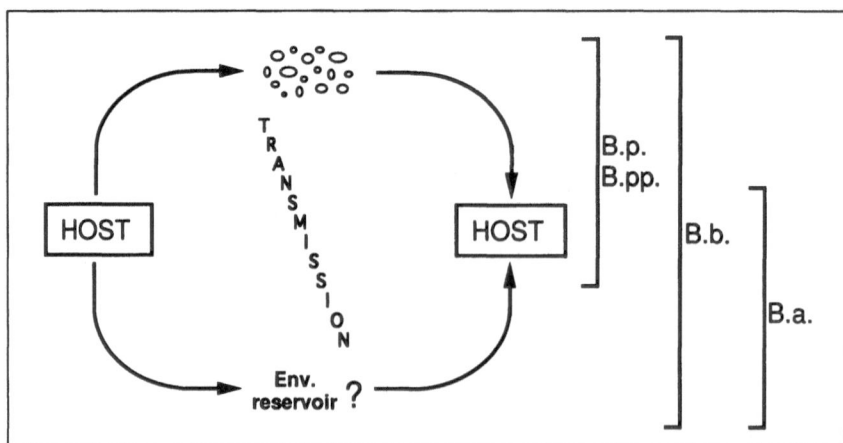

Fig. 2.5. A schematic representation of the life cycles of B. pertussis (B.p.), B. parapertussis (B.pp.), B. bronchiseptica (B.b.) and B. avium (B.a.). B. pertussis and B. parapertussis are thought to be transmitted directly from host to host by aerosol droplet. B. bronchiseptica is capable of surviving nutrient limiting conditions and may survive in an environmental reservoir. The postulated modes of transmission of B. bronchiseptica include aerosol droplets and environmental fomites. In contrast, B. avium has been suggested to transmit primarily by environmental mechanisms. See text for details.

The functional conservation of the BvgAS signal transduction system within and across *Bordetella* species suggests that it mediates an important adaptive response in these organisms. However, at what point in the *Bordetella* life cycle BvgAS sensory transduction is required, and what advantage phenotypic modulation confers on the organism is unknown. It is clear that the Bvg+ phase is required for virulence and the importance of several Bvg+ phase factors during infection has been established. A role for the Bvg− phase has not been demonstrated, but since *B. pertussis* is thought not to persist outside the host, it has been assumed that phenotypic modulation occurs in vivo. We have recently addressed the role of BvgAS sensory transduction in pathogenesis by constructing *Bordetella* strains with mutations that specifically affect signal transduction and comparing them with wild type for their ability to cause respiratory tract infection.[57] Since *B. pertussis* infects only humans, a model was developed based on the interaction between *B. bronchiseptica* and one of its natural hosts, the rabbit. When Bvg+ and Bvg− phase locked derivatives of a natural isolate of *B. bronchiseptica* were used to innoculate *Bordetella*-free rabbits, we were surprised to find that the Bvg+ phase locked mutant was indistinguishable from wild type (Fig. 2.6). The Bvg+ phase locked strain neither impaired nor enhanced in its ability to establish infection, colonize the entire respiratory tract and persist for up to 50 days despite the production of high titers of *Bordetella*-specific antibodies. Additionally, antibodies against Bvg− phase specific factors, including flagellin, were not detected. These results indicate that the Bvg+ phase is both necessary and sufficient for establishment of respiratory tract infection and strongly suggest that the Bvg− phase is not even expressed in vivo.

Since it appears that the Bvg− phase is not required for colonization, a role in transmission was proposed. Transmission can be divided into three steps; release of the organism from an infected host,

survival between hosts, and the initial interaction with the new host. To test whether phenotypic modulation is important for the initial interaction of the organism with a naive host, a rat model was used. The wild type and Bvg+ phase locked mutant strains described above grown under modulating and nonmodulating conditions were compared for their ability to establish infection by intranasal inoculation. All strains were equally capable of infecting rats with a similar, low ID_{50} (approximately 20 cfu), indicating that Bvg− phase factors are not required for this step (Akerley BA, Cotter PA, Miller JF, unpublished data).

B. bronchiseptica has been shown to be capable of surviving and multiplying in lake water and PBS.[8,9] These nutrient poor conditions could represent those encountered outside the host. Comparison of wild type and Bvg phase locked mutants indicated that only Bvg− phase organisms, induced either by mutation or by the presence of $MgSO_4$, could multiply in PBS. This result demonstrated a clear advantage for the Bvg− phase under nutrient limiting conditions,[7] and suggest a role for the Bvg− phase, and therefore for BvgAS signal transduction, in survival between hosts.

Taken together, the results described above suggest that the ability to switch from the Bvg+ phase to the Bvg− phase is not required during infection. Recently, another important aspect of BvgAS mediated signal transduction, the role of BvgAS mediated repression during pathogenesis, has recently been addressed.[111] By simultaneously mediating activation of virulence factors and repression of Bvg− phase specific factors, the BvgAS system prevents the expression of phenotypes specific to both phases at the same time. To investigate the importance of repressing the prominent Bvg− phase specific phenotype of motility during infection, a strain in which motility was expressed ectopically in the Bvg+ phase was constructed. This mutant was defective for colonization of the trachea in both rab-

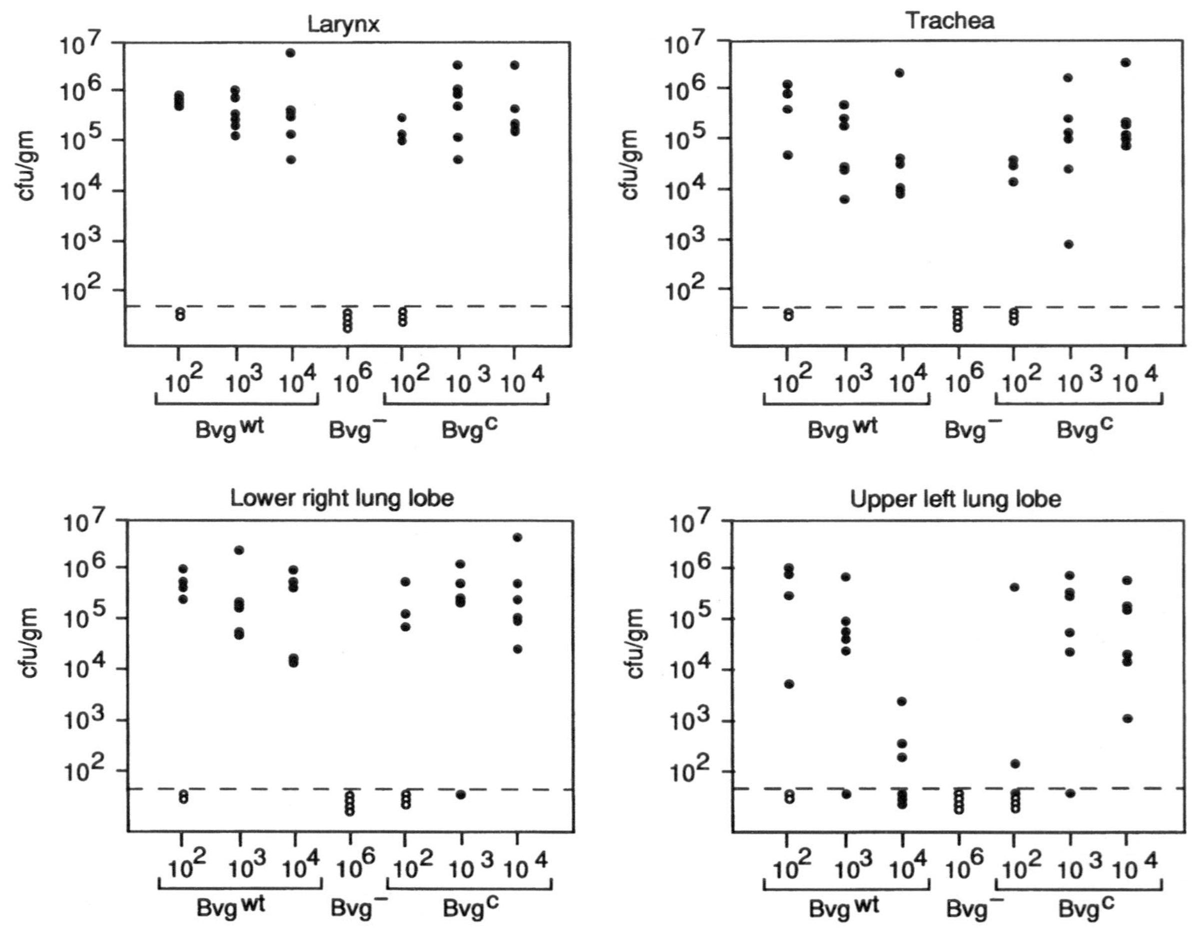

Fig. 2.6. Scatter plots showing colonization levels in the trachea, larynx, upper left and lower right lung lobes in individual female, 4 month old, New Zealand White rabbits inoculated intranasally with B. bronchiseptica strains. Rabbits were inoculated with the wild type parental strain (Bvg^wt), or with derivatives carrying an in frame deletion in bvgS (Bvg^-) or the bvgS-C3 mutation which confers a Bvg constitutive (Bvg^c) phenotype. Animals were sacrificed 3 weeks after inoculation. Bacterial counts expressed as cfu per gram of tissue are shown. Open circles represent uncolonized animals and solid circles represent infected rabbits. The lower limit of detection was 50 bacteria g⁻¹ as represented by dashed lines. Reprinted with permission from Cotter and Miller, Infect Immun 1994; 62:3381-3390.

bits and rats, demonstrating the importance of negative regulation by BvgAS during infection.[111]

The results described above demonstrate the importance of the Bvg[+] phase, characterized by expression of Bvg[+] phase factors as well as lack of expression of Bvg[−] phase factors, in the establishment of respiratory tract infection by *B. bronchiseptica*. The Bvg[−] phase was shown to be advantageous under nutrient limiting conditions and may be required for survival of the organism in the environment between hosts. Thus, it appears that a role for BvgAS mediated signal transduction is to sense whether the organism is within or outside of an animal host. Although these studies have been done with *B. bronchiseptica*, the functional conservation of BvgAS across species suggests these results may be extrapolated to *B. pertussis*. However, it seems unlikely that a change in gene expression pattern would be important for *B. pertussis* during the brief time that it's thought to exist outside of the human host. Additionally, Beattie et al have reported that a *vrg6*-Tn*phoA* mutant strain of *B. pertussis* was decreased in its ability to proliferate in the trachea and lungs of infected mice compared with wild type *B. pertussis*[58] suggesting a role for at least one Bvg repressed factor in vivo. Resolution of the apparent contrast between these results and those described above will require a careful comparison of the Bvg[−] phases of *B. pertussis* and *B. bronchiseptica* and a thorough assessment of the roles of each in the biology of the respective organisms. Despite the wealth of information that has accumulated, the role of BvgAS mediated phenotypic modulation in pathogenesis remains a mystery.

REFERENCES

1. The UNICEF Report. The state of the world's children 1994. U.S. Committee for UNICEF. New York, N.Y.

2. Gordon JE, Hood RI. Whooping cough and its epidemiological anomalies. Am J Med Sci 1951; 222:333-361.

3. Kendrick PL. Secondary familial attack rates from pertussis in vaccinated and unvaccinated children. Am J Hyg 1940; 32:89-91.

4. Lambert H J. Epidemiology of a small pertussis outbreak in Kent, Michigan. Public Health Report. 1965; 80:365.

5. Thomas MG, Redhead K, Lambert HP. Human serum antibody responses to *Bordetella pertussis* infection and pertussis vaccination. J Inf Dis 1989; 159:211-218.

6. Goodnow RA. Biology of *Bordetella bronchiseptica*. Microbiol Rev 1980; 44:722-738.

7. Cotter PA and Miller JF. BvgAS mediated signal transduction: Analysis of phase-locked regulatory mutants of *Bordetella bronchiseptica* in a rabbit model. Infect Immun 1994; 62: 3381-3390.

8. Porter JF, Parton R, Wardlaw AC. Growth and survival of *Bordetella bronchiseptica* in natural waters and in buffered saline without added nutrients. Appl Env Microbiol 1991; 57:1202-1206.

9. Porter JF, Wardlaw AC. Long-term survival of *Bordetella bronchiseptica* in lakewater and in buffered saline without added nutrients. FEMS Microbiol Lett 1993; 110:33-36.

10. Arp LH, Leyh RD, Griffith RW. Adherence of *Bordetella avium* to tracheal mucosa of turkeys: correlation with hemagglutination. Am J Vet Res 1988; 49:693-696.

11. Gray JG, Roberts JF, Dillman RC et al. Pathogensis of change in the upper respiratory tracts of turkeys experimentally infected with an *Alcaligenes faecalis* isolate. Infect Immun 1983; 42:350-355.

12. Saif YM, Moorhead PD, Dearth RN et al. Observations on *Alcaligenes faecalis* infection in turkeys. Avian Dis 1980; 24:665-684.

13. Simmons DG. Turkey coryza. In: Hofstad MS, Barnes HJ, Calnek BW, Reid WM, Yoder HW Jr, eds. Diseases of Poultry. 8. Ames, Iowa: Iowa State University Press 1984: 251-256.

14. Barnes HJ, Hofstad MS. Susceptibility of turkey poults from vaccinated and unvacinated hens to alcaligenes rhinotracheitis

(turkey coryza). Avian Dis 1983; 27:378-392.

15. Cimiotti W, Glunder G, Hinz KH. Survival of the bacterial turkey coryza agent. Vet Rec 1982; 110:304-306.

16. Simmons DG, Gray JG. Transmission of acute respiratory disease (rhinotracheitis) of turkeys. Avian Dis 1978; 23:132-138

17. Pittman M. Pertussis toxin: The cause of the harmful effects and prolonged immunity of whooping cough: A hypothesis. Rev Infect Dis 1979; 1:401-412.

18. Linnemann CC Jr, Nasenbury J. Pertussis in the adult. Ann Rev Med 1977; 28:179-185.

19. Tuomanen E, Weiss AA. Characterization of two adhesins of *Bordetella pertussis* for human ciliated respiratory epithelial cells. J Infect Dis 1985; 152:118-125.

20. Tuomanen E, Towbin H, Rosenfelder G et al. Receptor analogs and monoclonal antibodies that inhibit adherence of *Bordetella pertussis* for human ciliated respiratory epithelial cells. J Exp Med 1988; 168:267-277.

21. Relman D, Tuomanen E, Falkow S et al. Recognition of a bacterial adhesin by an integrin: Macrophage CR3 ($\alpha_M\beta_2$, CD11b/CD18) binds filamentous hemagglutinin of *Bordetella pertussis*. Cell. 1990; 61:1375-1382.

22. Roberts M, Fairweather NF, Leininger E et al. Construction and characterization of *Bordetella pertussis* mutants lacking the *vir*-regulated P.69 outer membrane protein. Mol Microbiol 1991; 5:1393-1404.

23. Leininger E, Roberts M, Kenimer JG et al. Pertactin, an Arg-Gly-Asp-containing *Bordetella pertussis* surface protein that promotes adherence of mammalian cells. Proc Natl Acad Sci USA 1991; 88:345-349.

24. Mooi FR, Jansen WH, Brunings H et al. Construction and analysis of *Bordetella pertussis* mutants defective in the production of fimbriae. Microbiol Path 1992; 12:127-135.

25. Goldman WE, Klapper DG, Basemen JB. Detection, isolation, and analysis of a released *Bordetella pertussis* product toxic to cultured tracheal cells. Infect

Immun 1982; 36:782-794.

26. Luker KE, Collier JL, Kolodziej EW et al. *Bordetella pertussis* tracheal cytotoxin and other muramyl peptides: Distinct structure-activity relationships for respiratory epithelial cytopathology. Proc Natl Acad Sci USA 1993; 90:2365-2369.

27. Endoh M, Takezawa T, Nakase Y. Adenylate cyclase activity of *Bordetella* organisms. Its production in liquid medium. Microbiol Immunol 1980; 24:95-104.

28. Confer DL, Eaton JW. Phagocyte impotence caused by an invasive bacterial adenylate cyclase. Science. 1982;217:948-950.

29. Hewlett EL, Gordon VM. Adenylate cyclase toxin of *Bordetella pertussis*. In: Wardlaw A, Parton R, eds. Pathogenesis and Immunity in Pertussis. New York: Wiley and Sons, 1988:193-209.

30. Irons LI, Gorringe AR. Pertussis toxin: production, purification, molecular structure, and assay. In: Wardlaw A, Parton R, eds. Pathogenesis and Immunity in Pertussis. New York: Wiley and Sons, 1988:95-120.

31. Masure H, Shattuck R, Storm D. Mechanisms of bacterial pathogenicity that involve production of calmodulin-sensitive adenylate cyclases. Microbiol Rev 1987; 51:60-65.

32. Ui M. The multiple biological activities of pertussis toxin. In: Wardlaw A, Parton R, eds. Pathogenesis and Immunity in Pertussis. New York: Wiley and Sons, 1988:121-146.

33. Munoz JJ, Bergman RK. *Bordetella pertussis*. Immunological and other biological activities. In: Rose N, ed. Immunology, Vol. 4. New York: Marcel Dekker, 1977:71-122,194-197

34. Lee CK, Roberts AL, Finn TM et al. A new assay for invasion of HeLa 229 cells by *Bordetella pertussis*: effects of inhibitors, phenotypic modulation and genetic alterations. Infect Immun 1990; 58:2516-2522.

35. Friedman RL, Nordensson K, Wilson L et al. Uptake and intracellular survival of *Bordetella pertussis* in human macrophages. Infect Immun 1992; 60:4578-4585.

36. Ewanowich CA, Melton AR, Weiss AA et al. Invasion of HeLa cells by virulent *Bordetella pertussis*. Infect Immun 1989; 57:2698-2704.

37. Saukkonen K, Cabellos C, Burroughs M et al. Integrin-mediated localization of *Bordetella pertussis* within macrophages: role in pulmonary colonization. J Exp Med 1991; 173:1143-1149.

38. Roop RM, Veit HP, Sinsky RJ et al. Virulence factors of *Bordetella bronchiseptica* associated with the production of infectious atrophic rhinitis and pneumonia in experimentally infected neonatal swine. Infect Immun 1987; 55:217-222.

39. Pittman M. In: Krieg NR, Holt JG, eds. Bergey's manual of systemic bacteriology. Vol. 1. London: Williams and Wilkins, 1984:388-393.

40. Aricò B, Gross R, Smida J et al. Evolutionary relationships in the genus *Bordetella*. Mol Microbiol 1987; 1:301-308.

42. Gentry-Weeks CR, Cookson BT, Goldman WE et al. Dermonecrotic toxin and tracheal cytotoxin, putative virulence factors of *Bordetella avium*. Infect Immun 1988; 56:1698-1707.

43. Jackwood MW, Saif YM. Pili of *Bordetella avium*: expression, characterization and role in in vitro adherence. Avian Dis 1987; 31:277-286.

44. Mooi FR, van der Heide HGJ, ter Avest AR et al. Characterization of fimbrial subunits from *Bordetella* species. Microbiol Path 1987; 2:473-484.

45. Simmons DG, Rose LP, Brogden KA et al. Partial characterization of the hemagglutinin of *Alcaligenes faecalis*. Avian Dis 1984; 28:700-709.

46. Miller JF, Mekalanos JJ, Falkow S. Coordinate regulation and sensory transduction in the control of bacterial virulence. Science 1989; 243:916-922.

47. Leslie PH, Gardner AD. The phases of *Haemophilus pertussis*. J Hyg 1931; 31:423-434.

48. Lacey BW. Antigenic modulation of *Bordetella pertussis*. J Hyg 1960; 58:57-93.

49. Melton AR, Weiss AA. Environmental regulation of expression of virulence determinants in *Bordetella pertussis*. J Bacteriol 1989; 171:6206-6212.

50. Weiss AA, Falkow S. Genetic analysis of phase variation in *Bordetella pertussis*. Infect Immun 1984; 43:263-269.

51. Aricò B, Scarlato V, Monack DM et al. Structural and genetic analysis of the *bvg* locus in *Bordetella* species. Mol Microbiol 1991; 5:2481-2491.

52. Stibitz S, Aaronson W, Monack D et al. Phase variation in *Bordetella pertussis* by a frameshift in a gene for a novel two component system. Nature. (London). 1989; 338:266-269.

53. Stibitz S, Yang MS. Subcellular localization and immunological detection of proteins encoded by the *vir* locus of *Bordetella pertussis*. J Bacteriol 1991; 173:4288-4296.

54. Parkinson JS, Kofoid EC. Communication modules in bacterial signalling proteins. Ann Rev Genet 1992; 26:71-112.

55. Stock J, Ninfa AJ, Stock AM. Protein phosphorylation and regulation of adaptive responses in bacteria. Microbiol Rev 1989; 53:450-490.

56. Beattie DT, Knapp S, Mekalanos JJ. Evidence that modulation requires sequences downstream of the promoters of two *vir*-repressed genes of *Bordetella pertussis*. J Bacteriol 1990; 172:6997-7004.

57. Knapp S, Mekalanos JJ. Two trans-acting regulatory genes (*vir* and *mod*) control antigenic modulation in *Bordetella pertussis*. J Bacteriol 1988; 170:5059-5066.

58. Beattie DT, Shahin R, Mekalanos JJ. A *vir*-repressed gene of *Bordetella pertussis* is required for virulence. Infect Immun 1992; 60:571-577.

59. Akerley BJ, Miller JF. Flagellin transcription in *Bordetella bronchiseptica* is regulated by the BvgAS virulence control system. J Bacteriol 1993; 175: 3468-3479.

60. Agiato Foster L-A, Giardina PC, Wang M et al. Siderophore biosynthesis in *Bordetella bronchiseptica* is controlled by the *bvg* regulon. American Society for Microbiology 93rd General Meeting,

Abstract 1558.

61. Miller JF, Roy CR, Falkow S. Analysis of *Bordetella pertussis* virulence gene regulation by use of transcriptional fusions in Escherichia coli. J Bacteriol 1989; 171: 6345-6348.

62. Miller JF, Johnson SA, Black WJ et al. Isolation and analysis of constitutive sensory transduction mutations in the *Bordetella pertussis bvgS* gene. J Bacteriol 1992; 174:970-979.

63. Aricò B, Miller JF, Roy C et al. Sequences required for expression of *Bordetella pertussis* virulence factors share homology with prokaryotic signal transduction proteins. Proc Natl Acad Sci USA 1989; 86:6671-6675.

64. Chang C, Winans SC. Functional roles assigned to the periplasmic, linker, and receiver domains of the *Agrobacterium tumefaciens* VirA protein. J Bacteriol 1992; 174:7033-7039.

65. Iuchi S, Lin ECC. Mutational analysis of signal transduction by ArcB, a membrane sensor protein for anaerobic repression of operons involved in the central aerobic pathways in *Escherichia coli*. J Bacteriol 1992; 174:3972-3980.

66. McCleary WR, Zusman DR. Purification and characterization of the *Myxococcus xanthus* FrzE protein shows that it has autophosphorylation activity. J Bacteriol 1990; 172:6661-6668.

67. Stibitz S. Complementation analysis of the *vir* locus of *Bordetella pertussis*. Submitted for publication. 1994.

68. Uhl MA, Miller JF. Autophosphorylation and phosphotransfer in the *Bordetella pertussis* BvgAS signal transduction cascade. Proc Natl Acad Sci USA 1994; 91:1163-1167.

69. Iuchi S. Phosphorylation/dephosphorylation of the receiver module at the conserved aspartate residue controls transphosphorylation activity of histidine kinase in sensor protein ArcB of *Escherichia coli*. J Biol Chem 1993; 268:23972-23980.

70. Kalman LV, Gunsalus RP. Nitrate- and molybdenum-independent signal transduction mutations in *narX* that alter regulation of anaerobic respiratory genes in *Escherichia coli*. J Bacteriol 1990; 172:7049-7056.

71. Collins LA, Egan SM, Stewart V. Mutational analysis reveals functional similarity between NarX, a nitrate sensor in *Escherichia coli* K-12, and the methyl-accepting chemotaxis proteins. J Bacteriol 1992; 174:3667-3675.

72. Ames P, Parkinson JS. Transmembrane signalling by bacterial chemoreceptors: *E. coli* transducers with locked signal input. Cell 1988; 55:817-826.

73. Roy CR, Miller JF, Falkow S. The *bvgA* gene of *Bordetella pertussis* encodes a transcriptional activator required for coordinate regulation of several virulence genes. J Bacteriol 1989; 171:6338-6344.

74. Boucher PE, Menozzi FD, Locht C. The modular architecture of bacterial response regulators: insights into the activation mechanism of the BvgA transactivator of *Bordetella pertussis*. J Mol Biol 1994; 241-377.

75. Swanson RV, Bourret RB, Melvin I. Simon. Intermolecular complementation of the kinase activity of CheA. Mol Microbiol 1993; 8:435-441.

76. Yang Y, Inouye M. Requirement of both kinase and phosphatase activities of an *Escherichia coli* receptor (Taz1) for ligands-dependent signal transduction. 1993; 231:335-342.

77. Genger JA, Dahlquist FW. Signal transduction in bacteria: CheW forms a reversible complex with the protein kinase CheA. Proc Natl Acad Sci USA 1991; 88:750-754.

78. Pan SQ, Charles T, Jin S et al. Preformed dimeric state of the sensor protein VirA is involved in plant-*Agrobacterium* signal transduction. Proc Natl Acad Sci USA 1993; 90:9939-9943.

79. Scarlato V, Prugnola A, Aricò B et al. Positive transcriptional feedback at the *bvg* locus controls expression of virulence factors in *Bordetella pertussis*. Proc Natl Acad Sci USA 1990; 87:6753-6757.

80. Aricò B, Rappuoli R. *Bordetella parapertussis* and *Bordetella bronchiseptica* contain transcriptionally silent pertussis toxin

genes. J Bacteriol 1987; 169:2847-2853.

81. Kloos WE, Mohapatra N, Dobrogosz WJ et al. Deoxyribonucleotide sequence relationships among *Bordetella* species. Int J Syst Bacteriol 1981; 31:173-176.

82. Musser JM, Hewlett EL, Peppler MS et al. Genetic diversity and relationships in populations of *Bordetella* spp. J Bacteriol 1986; 166:230-237.

83. Gentry-Weeks CR, Provence DL, J. M. Keith JM et al. Isolation and characterization of *Bordetella avium* phase variants. Infect Immun 1991; 59:4026-4033.

84. Akerley BJ, Monack DM, Falkow S et al. Role of the *bvgAS* locus in negative control of motility and flagella synthesis by *Bordetella bronchiseptica*. J Bacteriol 1992; 174:980-990.

85. Monack DM, Aricò B, Rappuoli R et al. Phase variants of *Bordetella bronchiseptica* arise by spontaneous deletions in the *vir* locus. Mol Microbiol 1989; 3:1719-1728.

86. Weiss AA, Hewlett EL. Virulence factors of *Bordetella pertussis*. Ann Rev Microbiol 1986; 40:661-686.

87. Charles IG, Dougan G, Pickard D et al. Molecular cloning and characterization of protective outer membrane protein P.69 from *Bordetella pertussis*. Proc Natl Acad Sci USA 1989; 86:3554-3558.

88. Stibitz S, Weiss AA, Falkow S. Genetic analysis of a region of the *Bordetella pertussis* chromosome encoding filamentous hemagglutinin and the pleiotropic regulatory locus *vir*. J Bacteriol 1988; 170:2904-2913.

89. Roy CR, Miller JF, Falkow S. Autogenous regulation of the *Bordetella pertussis* *bvgABC* operon. Proc Natl Acad Sci USA 1990; 87:3763-3767.

90. Roy CR, Falkow S. Identification of *Bordetella pertussis* regulatory sequences required for transcriptional activation of the *fhaB* gene and autoregulation of the *bvgAS* operon. J Bacteriol 1991; 173:2385-2392.

91. Miller JF, Roy CR, Falkow S. Regulation of *fhaB*, *bvg* and *ptx* transcription in *E. coli*: A comparative analysis. In Manclark CR ed: Proceedings of the Sixth International Symposium on Pertussis. Dept. of Health and Human Services, Bethesda, MD. pp217-224.

92. Locht C, Keith JM. Pertussis toxin gene: nucleotide sequence and genetic organization. Science 1986; 232:1258-1264.

93. Nicosia A, Perugini M, Franzini C et al. Cloning and sequencing of the pertussis toxin genes: operon structure and gene duplication. Proc Natl Acad Sci USA 1986; 83:4631-4635.

94. Gross R, Rappuoli R. Positive regulation of pertussis toxin expression. Proc Natl Acad Sci USA 1988; 85:3913-3917.

95. Gross R, Carbonetti NH, Rossi R et al. Functional analysis of the pertussis toxin promoter. Res Microbiol 1992; 143:671-681.

96. Carbonetti NH, Patamawenu A, Irish T et al. Differential regulation of virulence factor gene expression in *Bordetella pertussis*: effects of overexpression of RNA polymerase α subunit. American Society for Microbiology 93rd General Meeting, 1994; Abstract 1224.

97. Carbonetti NH, Khelef N, Guiso N, Gross R. A phase variant of *Bordetella pertussis* with a mutation in a new locus involved in the regulation of pertussis toxin and adenylate cyclase toxin expression. J Bacteriol 1993; 175:6679-6688.

98. Scarlato V, Aricò B, Prugnola A et al. DNA topology affects transcriptional regulation of the pertussis toxin gene of *Bordetella pertussis* in *Escherichia coli* and in vitro. J Bacteriol 1993; 175:4764-4771.

99. Glaser P, Sakamoto H, Bellalou J et al. Secretion of cyclolysin, the calmodulin-sensitive adenylate cyclase-hemolysin bifunctional protein of *Bordetella pertussis*. EMBO J 1988; 7:3997-4004.

100. Barry EM, Weiss AA, Ehrmann IE et al. *Bordetella pertussis* adenylate cyclase toxin and hemolytic activities require a second gene, *cyaA*, for activation. J Bacteriol 1991; 173:720-726.

101. Laoide BM, Ullmann A. Virulence dependent and independent regulation of the *Bordetella pertussis cya* operon. EMBO J 1990; 9:999-1005.

102. Goyard S, Ullmann A. Analysis of *Bordetella pertussis cya* operon regulation by use of *cya-lac* fusions. FEMS Microbiol Lett 1991; 77:251-256.

103. Livey I, Duggleby CJ, Robinson A. Cloning and nucleotide sequence analysis of the serotype 2 fimbrial subunit gene of *Bordetella pertussis*. Mol Microbiol 1987; 1:203-209.

104. Mooi FR, ter Avest A, van der Heide HGJ. Structure of the *Bordetella pertussis* gene encoding for the serotype 3 fimbrial subunit. FEMS Microbiol Lett 1990; 66:327-332.

105. Willems RJL, van der Heide HGJ, Mooi F. Characterization of a *Bordetella pertussis* fimbrial gene cluster which is located directly downstream of the filamentous hemagglutinin gene. Mol Microbiol 1992; 6:2661-2671.

105a. Willems R, Paul A, van der Heide HGJ et al. Fimbrial phase variation in *Bordetella pertussis*: a novel mechanism for transcriptional regulation. EMBO J 1990; 9:2803-2809.

106. Scarlato V, Aricò B, Prugnola A et al. Sequential activation and environmental regulation of virulence genes in *Bordetella pertussis*. EMBO J 1991; 10:3971-3975.

107. Rappuoli R, Aricò B, Scarlato V. Thermoregulation and reversible differentiation in *Bordetella*: a model for pathogenic bacteria. Mol Microbiol 1993; 6:2209-2211.

108. Melton AR, Weiss AA. Characterization of environmental regulators of *Bordetella pertussis*. Infect Immun 1993; 61:807-815.

109. Weiss AA, Melton AR, Walker KE et al. Use of the promoter fusion transposon Tn5*lac* to identify mutations in *Bordetella pertussis vir*-regulated genes. 1989; 57:2674-2682.

110. Beattie DT, Mahan MJ, Mekalanos JJ. Repressor binding to a regulatory site in the DNA coding sequence is sufficient to confer transcriptional regulation of the *vir*-repressed genes (*vrg* genes) in *Bordetella pertussis*. J Bacteriol 1993; 175:519-527.

111. Akerley BJ, Cotter PA, Miller JF. Ectopic expression of the flagellar regulon alters the development of the Bordetella-host interaction. Cell 1995; 80:611-620.

CHAPTER 3

SIGNAL TRANSDUCTION AND ENVIRONMENTAL STRESS IN CONTROL OF *PSEUDOMONAS AERUGINOSA* VIRULENCE

V. Deretic

ABSTRACT

Pseudomonas aeruginosa causes life threatening infections in several classes of patients with compromised natural defense systems. Among notorious examples are the systemic infections in severely burned patients and chronic respiratory infections in individuals with cystic fibrosis. Expression of virulence factors in this organism is controlled by several tiers of regulatory elements. Many of these regulatory factors belong to the superfamily of signal transduction systems originally termed bacterial two-component systems. For example, synthesis of the pili, two phospholipases, and the exopolysaccharide alginate are controlled by bona fide members of this superfamily of regulatory elements. In addition to these communication modules there are other sensory pathways which sometimes cooperate with the classical phosphotransfer response regulator/sensor systems. For instance, production of the alginate capsule-like protective coating and flagellin and pilin synthesis are controlled by alternative sigma factors. RNA polymerase σ subunits are frequently subject to control by negative regulators which suppress their activity unless growth conditions demand expression of their subordinate genes. In case of the alternative sigma factor agu (σ^E), which controls alginate production and certain other aspects of bacterial stress responses, a permanent loss of this negative regulation causes conversion to mucoid, alginate overproducing *P. aeruginosa* forms, frequently encountered in cystic fibrosis lung.

Signal Transduction and Bacterial Virulence, edited by Rino Rappuoli, Vincenzo Scarlato and Beatrice Aricò. © 1995 R.G. Landes Company.

1. INTRODUCTION

1.1. PSEUDOMONAS AERUGINOSA VIRULENCE FACTORS

Pseudomonas aeruginosa is traditionally classified as an opportunistic human pathogen.[1] This term should not be mistaken for a lower pathogenic potential. On the contrary, *P. aeruginosa* displays a respectful array of adhesins, powerful toxins, and other exoproducts. In several instances, its arsenal of virulence factors exceeds the known repertoire of some terminally differentiated human pathogens. The number and the range of function of *P. aeruginosa* virulence factors[2] clearly illustrates the pathogenic potential of this organism: (i) *P. aeruginosa* elaborates two ADP-ribosylating exoproducts, exotoxin A and exoenzyme S. Exotoxin A[3] is a bona fide toxin which ADP-ribosylates the eukaryotic elongation factor EF-2 at its modified histidine residue diphthamide. This is identical to the mode of action of diphtheria toxin. Both exotoxin A and diphtheria toxin share many commonalities including production under low iron conditions. Either case involves indirect or direct regulation by functionally analogous transcriptional factors responsive to iron availability. In the case of *P. aeruginosa* one such factor displays strong sequence similarity to the well studied ferric uptake repressor Fur of *Escherichia coli*.[4] The cell biology aspects (despite differences in the cell surface receptors to which they bind) of the endocytosis steps and the subsequent intracellular fate of the two toxins also appear similar, suggesting a high degree of specialization associated with this *P. aeruginosa* virulence factor. (ii) Pili act as the principal attachment factors.[5] (iii) Nonpilus adhesins, which remain to be fully characterized, mediated additional adherence to mucin and/or epithelial cells.[6] (iv) Elastases (LasA and LasB)[7,8] sinergistically degrade elastin, a major protein of the lung tissue. (v) Other proteases (e.g. alkaline protease) most likely contribute to tissue destruction.[9] (vi) Multiple siderophores (pyochelin and pyoverdin) and an extracellular redox cycling pigment (pyocyanin) together play a dual role in iron sequestration and generation of reactive oxygen intermediates (superoxide and hydroxyl radicals) capable of causing tissue damage.[10,11] (vii) LPS plays a role in serum resistance, antibiotic resistance, and as endotoxin in systemic infection.[12] (viii) Two phospholipases are capable of hydrolyzing the lung surfactant phosphatidylcholine and one of them has a hemolytic activity.[13] (ix) The exopolysaccharide alginate is overproduced during chronic infections of the cystic fibrosis lung and will be discussed extensively later in this chapter. (x) *P. aeruginosa* is also notorious for its innate antibiotic resistance.[14,15] (xi) Several other less studied factors are believed to enhance *P. aeruginosa* virulence.[2]

The robust panoply of *P. aeruginosa* pathogenic determinants, in combination with the lifestyle of this organism as a part-time pathogen, poses specific demands in terms of virulence gene expression. It is reasonable to expect that many of these traits must be regulated in order to be expressed under appropriate conditions encountered in the host. Furthermore, not all *P. aeruginosa* infections are alike. The course of events can be quite different, ranging from acute infections in burned patients to protracted chronic colonization of the cystic fibrosis lung. These extremes pose demanding logistics requirements with regard to the regulation of genes encoding different virulence factors. Some of these processes are illustrated in Figure 3.1. This chapter will cover those *P. aeruginosa* virulence factors that are regulated by signal transduction systems. We will also review how these elements integrate with the regulation by specific alternative sigma factors.

2. CONVERSION TO MUCOID FORMS

2.1. PSEUDOMONAS AERUGINOSA IN CYSTIC FIBROSIS

Considering the significant pathogenic potential of *P. aeruginosa*, it appears that

Fig. 3.1. Regulation of P. aeruginosa virulence factors by signal transduction elements, alternative sigma factors, and autoinduction systems. Asterisks denote response regulators from the superfamily of bacterial two-component systems.

(A) Expression of the flagellin gene is dependent on σ^{28} (FliA). The activity of this alternative sigma factor in S. typhimurium is under negative control by its cognate antisigma factor.

(B) Control of the pilin gene pilA. This gene is activated by the PilS-PilR sensor/response regulator system. The transcription of the pilA promoter is also dependent on the alternative sigma factor σ^{54} (RpoN). Nonpilus adhesins which interact with mucin and/or epithelial cells are not expressed in rpoN mutants.

(C) The lasB gene encodes an extracellular protease, elastase. It is under the control of a cell density responsive system consisting of the transcriptional activator LasR and the cell density signal generator LasI (the putative homoserine lactone synthase). The LasR and LasI are similar to the LuxR-LuxI system controlling cell-density responsive luminescence in Vibrio fischeri. PAI, P. aeruginosa autoinducer (a homoserine lactone derivative). LasR most likely directly interacts with the lasB gene which has a well conserved lux box at a typical location, -40 bp upstream of the lasB mRNA start site. LasR also affects expression of lasA and some other exoproducts, which act not only as virulence factors but can be found in immune complexes that contribute to the inflammation in the CF lung.

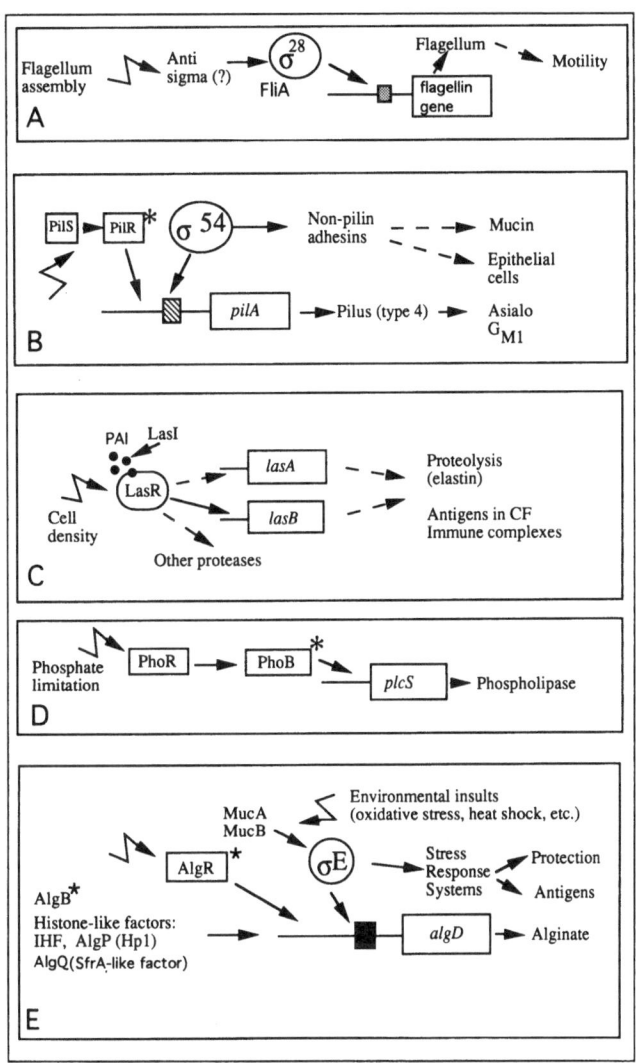

(D) The plcS gene is the structural gene for the hemolytic phospholipase PLC-H of Pseudomonas aeruginosa. This gene, and the gene encoding the nonhemolytic phospholipase (plcN) are controlled by the PhoR-PhoB two-component signal transduction system. A functionally interchangeable system controls the phosphate regulon in E. coli. In addition to phosphate limitation, expression of the plc genes in P. aeruginosa is affected by osmolarity and osmoprotectants.

(E) Overproduction of the exopolysaccharide alginate results in mucoid colony appearance and is considered to be a major pathogenic determinant expressed during chronic infections in cystic fibrosis. Several regulatory factors control expression of the critical biosynthetic gene promoter, algD. Two signal transduction elements, the response regulators AlgR and AlgB, affect algD expression. AlgR binds directly to three sites within the algD promoter and augments its transcription. Initiation of algD transcription is completely dependent upon the alternative factor σ^E. This sigma factor is normally inhibited by MucA and MucB. Environmental stress reduces or removes this negative regulation permitting induction of σ^E activity. In addition to alginate, σ^E controls other functions improving the ability of Pseudomonas to cope with heat shock, oxidative stress, etc. Other ancillary factors (histone-like elements IHF and AlgP) may affect the geometry of the P. aeruginosa nucleoid and algD activity. However, these effects are probably generic and structural in nature, and should not be considered as regulation. AlgQ has been erroneously reported as a kinase phosphorylating AlgR. Instead, this protein is homologous to and functionally interchangeable with SfrQ, a regulator of siderophore synthesis.

this ubiquitous and otherwise metabolically quite versatile organism did not undergo terminal specialization such as that seen in bacteria locked in the state of permanent association with the human or other hosts. However, *P. aeruginosa* is only a small step away from being capable of causing long-term colonization of humans. This is perhaps best illustrated in the case of cystic fibrosis (CF).[16] This most common inheritable disease among Caucasians (with incidence of approximately 1:2,500) is characterized by defective chloride transport across apical membranes of epithelial cells. This deficiency affects the function of almost all exocrinous glands. The most serious consequences of the genetic defect in CF[16] are associated with the lung pathology. The altered environment of the lung, which per se appears to be relatively mild, sufficiently perturbs the natural clearance mechanisms of the respiratory tract permitting recurring respiratory infections with different pathogens.[17] In most cases, the succession of pathogens usually ends in a permanent colonization with *P. aeruginosa*. This bacterium is capable of persisting in the CF respiratory tract for many years. Chronic colonization with *P. aeruginosa* and the associated inflammatory processes are the primary cause of high mortality and morbidity in CF.[17] It is partially due to the interest in this specific host-pathogen interaction that we have the current insight into the regulation of virulence factors in *P. aeruginosa*.

2.2. FUNCTION OF ALGINATE COATING

The typical mucoid colony morphology of *P. aeruginosa* isolates from CF patients is due to excessive production of the exopolysaccharide.[17,18] All *P. aeruginosa* strains, regardless of the source, have the genetic capacity to synthesize alginate. However, only in a few instances is it possible to isolate mucoid *P. aeruginosa* from other sites except the CF lungs.[18] This morphotype is so characteristic of CF strains that it is sometimes used as a patho-

gnomonic sign. Due to the paramount role that alginate plays in CF, we will summarize some of its features and refer the reader to a recent review[19] for the details of its chemical structure. Alginate is a linear copolymer of mannuronic and its C-5 epimer guluronic acids. It is partially acetylated. This chemical composition translates into the following physicochemical and biological properties: (i) Alginate is a highly negatively charged hydrophilic compound which when hydrated forms viscous solutions. Depending upon the concentration of divalent cations such as calcium, the physicochemical state of alginate solutions can change ranging from sols to rigid gels. (ii) The guluronate to mannuronate ratio affects the gelling properties of alginate, and may also have important effects on the ability to elicit inflammatory cytokines. For example, microcapsules composed of alginates enriched for guluronate blocks, when injected into peritoneal cavity, appear to induce less inflammatory cytokines IL-1 and TNFα than mannuronate rich alginates. (iii) Acetylation may affect the rate of conversion from mannuronate to guluronate. The presence of acetyl groups appears to play an important role in suppression of lymphocyte and neutrophil functions, and in scavenging reactive oxygen intermediates by alginate.[20,21] It is also likely that the presence of hydrophobic groups, provided by acetylation of this otherwise highly hydrophilic polymer, broadens the adhesion properties of alginate. (iv) *P. aeruginosa* growing in the CF lung inside alginate embedded microcolonies is protected from effective opsonization and phagocytosis, in a manner analogous to the protection of *Streptococcus pneumoniae* afforded by its capsule. In either case opsonins appear to penetrate the porous matrix of the hydrated exopolysaccharide and are deposited on the bacterial surface, but are inaccessible to their cognate receptors on phagocytic cells due to the presence of the physical barrier.[22] (v) A well documented but by and large less advertised characteristic of alginate is that it can scavenge reactive oxygen intermediates such as superoxide radicals and hypochlorite.[20,21] The significance of this property will become apparent when we address the environmental signals that may be transduced via regulatory systems which control alginate production and conversion to mucoidy.

2.3. ALGINATE BIOSYNTHETIC PATHWAY

The alginate biosynthetic pathway and chromosomal organization of the genes encoding biosynthetic enzymes are given in Figure 3.2. Starting from the pool of central sugar intermediates, fructose 6-phosphate is converted via a number of intermediates (mannose 6-phosphate, mannose 1-phosphate, and GDPmannose) into GDPmannuronic acid, a direct precursor for alginate polymerization (Fig. 3.2B). Most of the genes involved in this pathway are clustered at 34 minutes on the chromosome of *P. aeruginosa* (Fig. 3.2A). The genes and their functions are listed in the order of transcription within the cluster. (i) *algD* encodes GDPmannose dehydrogenase (GMDH)[23] which catalyzes double oxidation of GDPmannose into GDPmannuronic acid, representing the final stage in alginate precursor synthesis. (ii) The *algG* gene encodes the epimerase which converts mannuronic acid into guluronic residues.[24] (iii) The gene product of *algL* has an alginate depolymerase activity.[25] (iv) The *algF* gene encodes an acetylase.[26] (v) *algA*, encodes an enzyme with phosphomannose isomerase and a pyrophosphorylase activity and may participate in two non-consecutive steps of the alginate biosynthetic pathway.[27] There are also other genes in this cluster for which the precise function has not been established but it is likely that some of them play a role in alginate polymerization and export. Conversely, a well characterized alginate gene, *algC*, is excluded from the cluster and maps at a different location on the chromosome at 10 minutes. At first, this may appear as a puzzling complication, but it most likely re-

flects the participation of the *algC* gene product, that has both phosphommanomutase and phosphoglucomutase activities, in the synthesis of the LPS core in addition to its role in alginate biogenesis (Fig. 3.2B).[28,29]

2.4. TRANSCRIPTIONAL ACTIVATION OF *ALGD* IN MUCOID CELLS: ALGINATE VIS-Á-VIS LPS

The *algD* gene is activated in mucoid cells. The gene product of *algD*, GMDH, catalyzes a 4-electron transfer, NAD-dependent oxidation of GDPmannose into GDPmannuronate. This process appears to be a "thermodynamical switch" (Fig. 3.2B). This reaction, that for all practical purposes can be considered irreversible, channels sugar intermediates into the alginate biosynthetic pathway. Why is such a pivotal role given to the overproduction of an enzyme positioned so late in the biosynthetic pathway? Recent analyses of the shared steps in the synthesis of alginate and LPS, suggest that the enzymes catalyzing the first three steps of alginate precursor biosynthesis also participate in the synthesis of the LPS core and polysaccharide side chains (Fig. 3.2B).[28,29] Thus, *algD* is the first alginate specific gene, and its activation represents the first committing step for alginate synthesis.

The *algD* gene also has a strategic location at the beginning of the alginate gene cluster (Fig. 3.2A). Its promoter is activated in mucoid cells to high levels, but appears virtually silent in laboratory grown nonmucoid cells. It is likely that the activation of the *algD* promoter affects transcription of other genes located further downstream in the cluster.[30] Thus, a key event associated with the mucoid status of the cell is the transcriptional activation of *algD*.[23]

2.5. RESPONSE REGULATORS IN THE CONTROL OF ALGINATE GENE EXPRESSION

The *algD* promoter has become the roseta stone for uncovering the molecular mechanism governing expression of

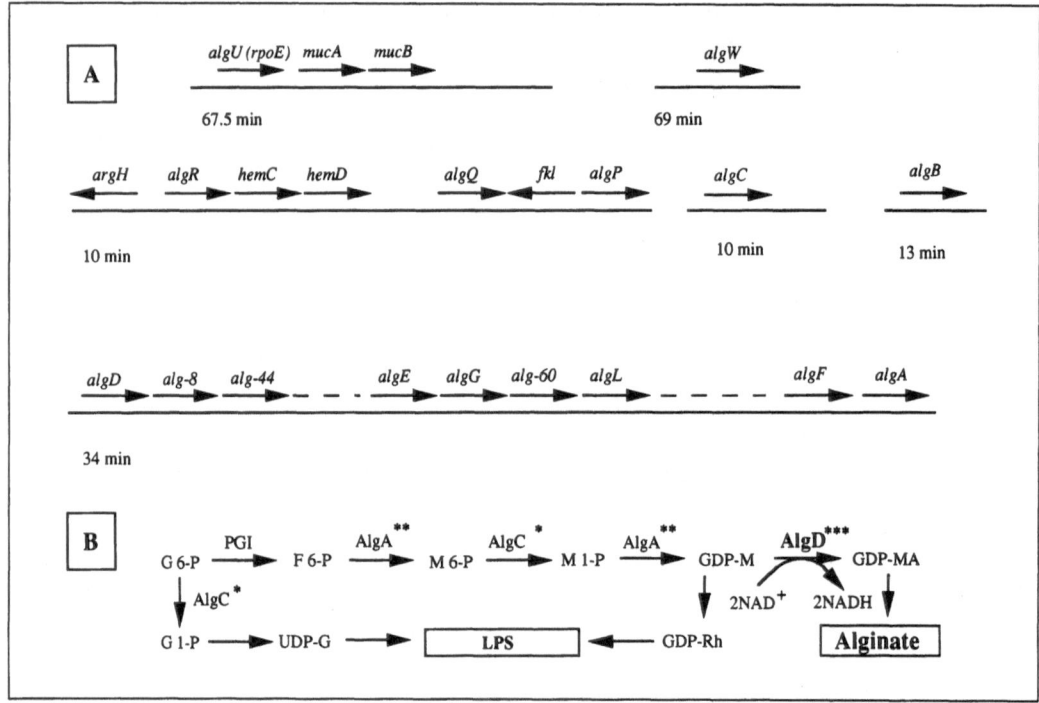

Fig. 3.2. Alginate genes and biosynthetic pathway.

(A) At least six chromosomal loci are involved in the synthesis of alginate and its regulation. The cluster of three genes at 67.5 minutes encodes the alternative sigma factor AlgU, which displays strong similarity with and is functionally interchangeable with the E. coli σ^E, and its negative regulators mucA and mucB. AlgU initiates transcription of the algD promoter (algD encodes the first alginate specific enzyme) and augments its own transcription and that of the response regulator algR. Mutations which inactivate mucA cause conversion to mucoidy in a significant fraction of P. aeruginosa isolates from cystic fibrosis (Fig. 3.5). AlgU is another gene that affects conversion to mucoidy. The response regulators algR and algB also affect algD transcription. The algR gene is linked to the auxotrophic marker argH and is co-transcribed in mucoid cells with the hemC and hemD genes involved in heme biosynthesis. Because algD is at the beginning of the alginate biosynthetic gene cluster, activation of the algD promoter enhances expression of the genes located further downstream. The genes with the known function in this cluster are: GDPmannose dehydrogenase; algG, epimerase; algL, alginate lyase; algF, acetylase; and algA, phosphomannose isomerase/GDPmannose pyrophosphorylase (a bifunctional enzyme catalyzing two nonconsecutive steps in the precursor synthesis). The algC gene maps in the same general location as algR (10 minutes; however, it is not closely linked to the algR-algQ-algP region). This gene encodes phosphoglucomutase/phosphomannomutase which participates in the generation of glucose 1-phosphate and GDPmannose, both of which are required for the synthesis of LPS while only the latter is needed for alginate synthesis (see panel B). algQ and algP have been discussed in the legend to Fig. 3.1. fkl, is an FK506 binding protein-like factor. It shows strong homology to Mip of Legionella pneumophila. Its role in mucoidy has not been established.

(B) Relationship of the alginate and LPS biosynthetic pathways (see text). G 6-P, glucose 6-phosphate; F 6-P, fructose 6-phosphate; M 6-P, mannose 6-phosphate; M 1-P, mannose 1-phosphate; GDP-M, GDPmannose; GDP-MA, GDPmannuronic acid; G 1-P, glucose 1-phosphate; UDP-G, UDP glucose; GDP-Rh, GDPrhamnose. PGI, phosphoglucoisomerase; AlgD, GDPmannose dehydrogenase which catalyzes the committing step in alginate synthesis (a double oxidation of GDPmannose coupled to reduction of two molecules of NAD⁺). AlgA and AlgC are described under A. Single or double asterisk, enzymes shared between alginate and LPS biosynthetic pathways. Triple asterisks, first alginate specific enzyme.

mucoidy in *P. aeruginosa.*[31] These investigations indicated early on that signal transduction elements are at the heart of the regulation of *algD.* Two response regulators, AlgR[32] and AlgB[33], have been implicated in the control of *algD.* It is perhaps worth mentioning that *algR* is also one of the first bacterial response regulators reported to control a virulent trait.[32] This gene has been isolated and sequenced at the time when similar regulatory elements have been recognized in the control of virulence in *Bordetella pertussis, Salmonella typhimurium, Vibrio cholerae,* and *Staphylococcus aureus.*[34] It is thus easy to understand the excitement that this finding has brought to the investigators in the field, since it has suggested that environmental signals, perhaps related to those encountered in the CF lung, may be responsible for conversion to mucoidy. Although these processes and the signals are still not understood, AlgR remains one of the most interesting and best characterized components in the regulation of alginate synthesis.

2.6. ALGR: BIOCHEMICAL PROPERTIES, PHOSPHORYLATION, AND BINDING TO *ALGD*

AlgR binds directly to the *algD* promoter and is absolutely necessary for the expression of this gene at levels required to attain mucoid phenotype. AlgR has been purified to near homogeneity and its N-terminal sequence determined, as well as that of several internal peptides,[35] confirming its primary structure predicted based on the DNA sequence. AlgR is a 27.6 kDa polypeptide, which has a typical regulator N-terminal domain of close to 120 amino acids, with highly conserved Asp residues at position 9 and 54 and Lys at position 102, corresponding to Asp13, Asp57 and Lys109 of CheY, the response regulator usually used as a reference. Molecular modeling of the N-terminal domain of AlgR (Fig. 3.3A) suggests a very good overlap with the amino acid backbone of CheY, for which the crystal structure is known.[36] This is noticeable in particular in the regions which comprise the conserved hydrophobic core and the critical residues within the acidic pocket of this molecule (Fig. 3.3B).[36] Biochemical analyses[37] have indicated that AlgR can undergo a typical phosphotransfer reaction when incubated with a phosphorylated prototype kinase/sensor molecule such as CheA. This phosphotransfer depends on active AlgR and takes place even if CheA is heat inactivated.[37] Also, chemically synthesized phosphoramidates appear to be phosphodonors in reaction with AlgR.[37] These features suggest that the catalytic ability for the phosphotransfer lies within AlgR. This protein can also interact with small molecular weight compounds such as acetyl phosphate and carbamoyl phosphate,[37] which are participants in the intermediary metabolism. Interestingly, an AlgR-cognate protein/sensor histidine kinase, which presumably senses the environmental signals and transduces them to AlgR, has not been identified. The role of AlgQ,[38] formerly a candidate for the sensor, is unclear. This protein has no similarities with the orthodox histidine protein kinases and instead is similar and functionally interchangeable with PrfA, a positive regulator of sederophore biosynthesis.[39] Although the formal possibility that there is a specific protein histidine kinase interacting with AlgR cannot be ruled out, the alternative that AlgR interacts in vivo with several proteins and perhaps with small molecular weight phosphodonors is equally plausible.

Purified AlgR is a monomer. Interestingly, it has a tendency to aggregate once the salt concentration falls below 100 mM.[35] Its interactions with the critical target promoter *algD* have been studied in detail. AlgR bound to DNA protects approximately 18-22 bp from DNaseI digestion. Centered within this footprint is the AlgR recognition sequence ACCGTTGTC, termed core.[35,40] The core sequence has been defined by a relatively tight consensus found in different AlgR-binding sites and by mutagenesis of individual nucleotides

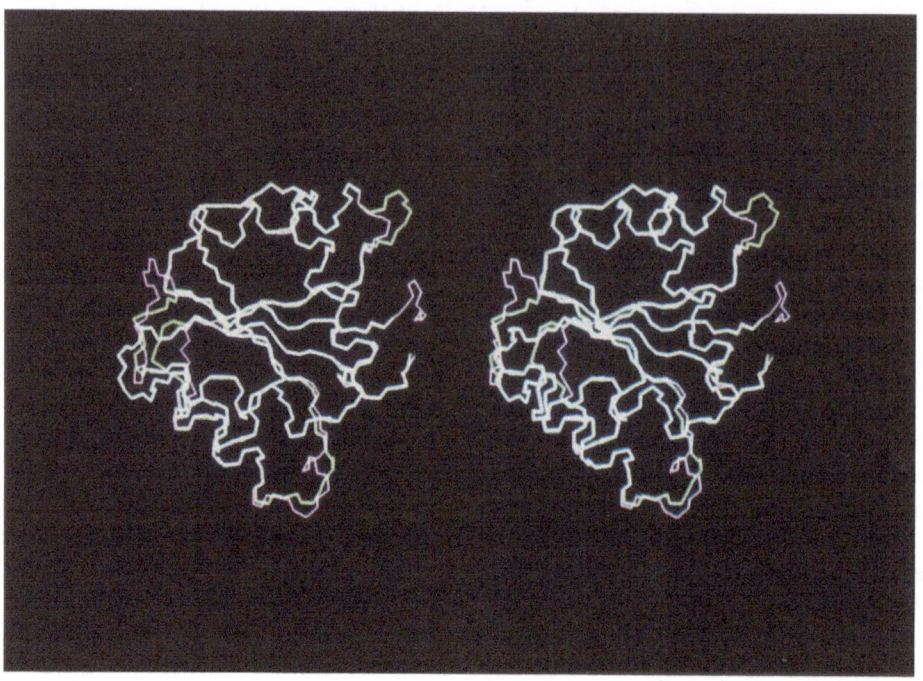

Fig. 3.3A

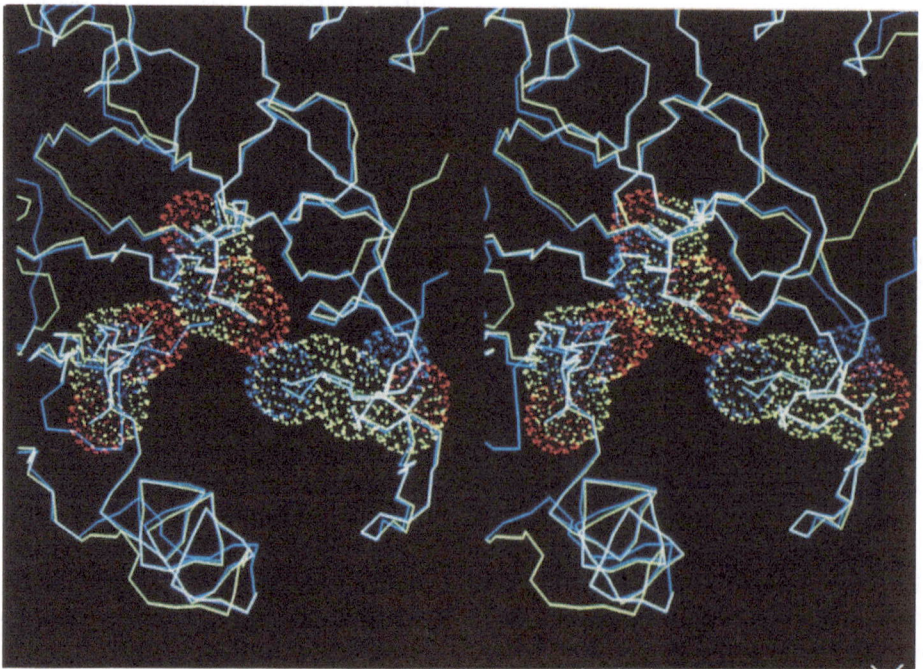

Fig. 3.3B

within such sites.[40] The absence of palindromic sequences within AlgR binding sites is consistent with AlgR being a monomer. There are three AlgR binding sites within the *algD* promoter: RB1, RB2, and RB3, each containing a single recognition sequence. These sites are separated by long intervening sequences of various lengths. The two high affinity binding sites, RB1 and RB2, located far upstream from the mRNA start site with the core sequences ending at -463 and -387 relative to the *algD* mRNA start site, respectively, have approximately 1 kcal/mol (4.18 kJoul/mol) higher energy of binding than the third site RB3. Interestingly, this difference in free energy of binding is similar to the difference in free energy of binding of bacteriophage λ cI repressor to the operator sites O_L1 and O_L3 (-1.4 kcal/mol [-5.8 kJ/mol]). The weaker RB3 site is juxtaposed to the RNA polymerase binding site, and this proximity probably makes up for the lower affinity.[40] Each site contributes to the activation of *algD* and is required to attain expression levels compatible with the mucoid phenotype. It is possible that AlgR molecules bound at different sites interact with each other, which could be of significance for the loading of the low affinity site RB3.[40] Several histone-like [Hp1 (also known as AlgP) and IHF] have been implicated in promoting such interactions or interactions of AlgR with the

RNA polymerase.[31] Whether or not AlgR phosphorylation modulates interactions between AlgR molecules bound at different sites, or how it affects other aspects of transcriptional activation events is currently not known. We have recently generated mutants of AlgR with altered residues Asp9 and Asp54 (corresponding to the Asp13, and Asp57 of CheY), and purified these proteins. These mutant forms of AlgR lose their ability to be phosphorylated in reaction with CheA-phosphate in vitro but clearly retain their ability to bind individual sites such as RB2, without significant reduction in affinity as compared to the wild type AlgR (Krieg DP, Hibler NS, Deretic V; unpublished data). It remains to be determined how these mutations alter AlgR function and at what level may phosphorylation be important.

2.7 ALGR AND ENVIRONMENTAL CONTROL OF *ALGD*

In addition to the missing cognate kinase, the environmental signals recognized by AlgR remain unknown. It has been proposed that AlgR may act as an osmosensor.[41] This has been based on the behavior of some strains that can increase alginate production in the presence of salt concentrations.[41] Despite the certain teleological appeal of this hypothesis due to the increased salt concentrations in some CF secretions, this theory does not hold

Fig. 3.3 (opposite). (A) Stereo image of the N-terminal domain of AlgR modeled after CheY (Hart J, Robertus J, Deretic V; unpublished data). Computer modeling was done by substituting the CheY crystal structure coordinates with AlgR residues. Energy minimization runs were performed with charges turned off (100 cycles) followed by 100 cycles with charges turned on. The light blue line is the amino acid backbone where according to the model AlgR and CheY show complete overlap. Purple, CheY backbone that folds differently from AlgR; green, AlgR regions that according to the computer model fold differently from CheY.

(B) A close-up view of the acidic pocket (the putative active site) of the AlgR model in A. Sticks and dots rendering of the side chains is given for the critical residues Asp9, Asp54, and Lys109 (AlgR numbering). These residues are located at the C-terminal edge of the β sheet at the base of the loops connecting the C-terminal endings of β strands with adjacent helices lying on the alternating sides of the central β1 sheet: Asp9 (β1/Asp9/ α1) is on the left side of the pocket; α1 helix is viewed along its long axes, going away from the observer. Asp54 (β3 /Asp54/α3) occupies the center of the pocket. Lys102 (β5/Lys102/α5) is on the right side of the pocket. Lys102 extends into the pocket by virtue of its cis-peptide bond with the following residue Pro103 based on the refined CheY crystal structure. The model predicts that, as in the case of CheY, the ε amino group of Lys102 may be in a close bonding contact with the carboxyl group of Asp54. Light blue regions, overlap between the predicted AlgR backbone and CheY. Dark blue, CheY; green, AlgR backbone different from CheY.

up against the scrutiny of thorough examinations. For one, the strains which increase alginate production in the presence of 0.3M NaCl represent a small minority of mucoid isolates, and should be viewed more as an exception than a rule. Furthermore, it should be emphasized that the studies examining these effects have been performed using mucoid mutants (see later sections on *muc* mutations), and thus the conclusion drawn from such experiments cannot be taken as faithfully representing the natural induction principles. Moreover, the observed induction is *algR*-independent. When *algR* is insertionally inactivated on the chromosome of strains which do respond to osmolarity changes, the induction ratios of *algD* transcription remain the same in both *algR*+ strains and in their isogenic *algR:Tc^r* derivatives.[42] To be sure, the overall level of *algD* transcription is dramatically reduced in *algR* mutants and the cells lose mucoid colony morphology as expected, but the osmolarity-dependent response of *algD* stays unaltered.[42] The nature of this AlgR-independent osmolarity induction on one side, and the real signals recognized by AlgR on the other side, both remain to be identified.

It is also worth mentioning in the context of potential signals recognized by this system, that *algR* is tightly linked to and cotranscribed with the *P. aeruginosa* *hemC* and *hemD* genes.[43] Mutations in *hemC* and *hemD* have a similar (although milder) effect on alginate induction as compared to *algR* mutations in the same strains.[43] The *hemC* and *hemD* genes encode porphobilinogen deaminase and uroporphyrinogen III cosynthase, respectively. These enzymes are critical for heme biosynthesis. Interestingly, mutations in the *hem* genes in *E. coli* mimic low oxygen pressure and affect expression of genes regulated by the *arcA-arcB* system, which is a two-component system controlling TCA cycle enzymes, terminal oxidases (cytochrome o and cytochrome d), and some primary dehydrogeneases in response to oxygen availability. Such findings suggest

that perhaps there may be a link between the redox state of the cell or availability of terminal electron acceptors (oxygen, nitrate) and the NAD-dependent double oxidation of GDPmannose catalyzed by the gene product *algD*. The hypothesis that oxygen availability may play a role in these processes has merit considering the fact that CF patients are often given prolonged oxygen inhalation therapy. Nevertheless, these considerations are merely suppositions at this stage, and a lot more needs to be learned about the precise signals transduced via AlgR.

2.8. OTHER RESPONSE REGULATORS AFFECTING ALGINATE SYNTHESIS

Another signal transduction element implicated in the control of alginate is *algB*.[33,44] This gene encodes an interesting response regulator that belongs to the RO$_{IV}$ subfamily of bacterial signal transducers.[45] In addition to a typical N-terminal domain, it has a central domain with the motif resembling nucleotide binding folds. The corresponding central domain in the well-studied homologue NtrC has ATPase activity and plays a role in the formation of open complexes during the process of RpoN (σ^{54}) dependent initiation of transcription.[46] AlgB has been purified, but its binding or any other direct interaction with the *algD* promoter has not been established. Thus, it remains to be seen whether the effects of mutations in *algB* are indirect, possibly reflecting disturbance in the flux of phosphotransfer processes, or this gene affects expression of other factors controlling *algD* or other aspect of alginate biosynthesis and processing.

2.9. MECHANISM OF CONVERSION TO MUCOID PHENOTYPE

The story of alginate control would not be complete without a brief description of the molecular mechanism of conversion to mucoidy (Fig. 3.4). After many years of investigation by numerous groups, it has transpired that the switch to mucoidy is based on mutations which allow

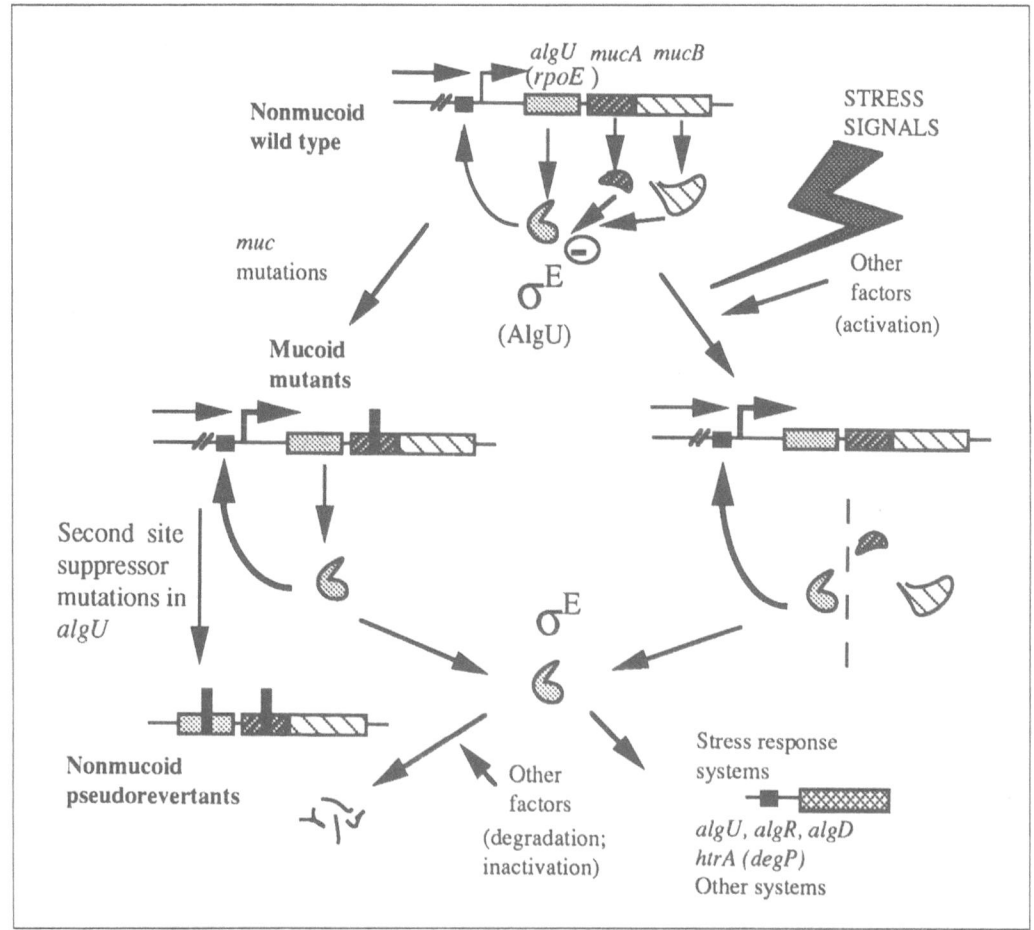

Fig. 3.4. Mechanism of conversion to mucoidy in Pseudomonas aeruginosa. *At the top of the regulatory hierarchy is the algU mucA mucB gene cluster. The gene product of algU is the alternative sigma factor σ^E, which in E. coli has been sometimes referred to as "the second heat shock sigma factor". This sigma factor controls systems which enhance bacterial ability to survive extreme stress conditions such as very high temperatures and exposure to reactive oxygen intermediates. AlgU (σ^E) is under negative control by at least two factors: MucA and MucB. The left side of the scheme depicts how mutations that inactivate mucA (vertical line in mucA) result in conversion to mucoid forms which constitutively produce alginate. Such mutants are frequently isolated from the lungs of patients with cystic fibrosis. In the absence of a functional MucA, σ^E is present in amounts sufficient to transcribe its subordinate promoters which include algD, algR, and the promoter of its own gene (algU is transcribed from several promoters as indicated by horizontal arrows; the proximal promoter is dependent on σ^E; bent arrow). The toxicity of σ^E, when it is present in excessive amounts and not being countered by its negative regulators, puts selective pressure on mucA P. aeruginosa mutants, causing them to "revert" to nonmucoid phenotype. This process is not a true reversion, and is based on various second site suppressor mutations. In some cases these occur in the algU gene giving rise to nonmucoid pseudorevertants. The right side of the scheme depicts the physiological induction of the system (in the absence of mutations). This system is most likely designed to transiently induce defense systems to help bacteria endure unfavorable environmental conditions. The processes underlying this phenomenon, and the range and identity of genes (other than alg) that are induced by σ^E activation (some of which are listed; e.g. htrA [degP]) is currently under investigation. .*

the alternative sigma factor σ^E (AlgU) to express its function. The mutations causing mucoidy in a significant number of CF isolates inactivate the negative regulator *mucA* which normally suppresses the function of σ^E (Fig. 3.4, left side). Once σ^E is relieved from this control, the *algD* promoter and several other promoters become active.[31,47-50] AlgR further stimulates this activity. One model that is currently being considered includes interactions of σ^E, RNA polymerase core, and AlgR, that together bring about physiological induction of the system (Fig. 3.4, right side) in the absence of *mucA* mutations.

σ^E has been defined in *E. coli* at the biochemical level as an alternative sigma factor associated with a minor RNA polymerase species.[51] The σ^E containing RNA polymerase holoenzyme transcribes one of the several promoters of the major heat shock sigma factor σ^{32} (RpoH, HtpR). This promoter is the only *rpoH* promoter active at temperatures close to 50°C. It is interesting that despite the characterization of σ^E at the protein level, no corresponding gene has been isolated until very recently. In fact it was the similarity (66% identity and 91% overall similarity) of the gene product of *algU* [this is how the gene encoding σ^E in *P. aeruginosa* has been initially termed[47]], and its analog from *E. coli* that led to the uncovering of the *rpoE* genes in *E. coli* and *S. typhimurium*.[31,50]

The σ^E RNA polymerase holoenzyme is also the only polymerase transcribing *htrA* (*degP*), a gene necessary for survival under conditions of extreme heat shock.[51] In *Salmonella typhimurium*, *htrA* is required for resistance to reactive oxygen intermediates such as hydroxyl and superoxide radicals and for the virulence of this organism in mice.[52] Our recent studies show that in addition to heat shock, osmotic stress can also induce at least one of the σ^E-dependent promoters (viz. *algR*) (Schurr MJ and Deretic V, unpublished data). There are several important ramifications of these findings: (i) Production of alginate is most likely an integral part of the

cellular stress response. (ii) In vivo selection of mutants with the system locked in the state of constitutive expression, such as in mucoid strains isolated from CF patients, suggests that overexpression of σ^E-dependent systems, including alginate, may be critical for the long-term survival of *P. aeruginosa* in the CF lung. (iii) Since *algU* (*rpoE*) mutants in *P. aeruginosa* show increased sensitivity to killing by superoxide generating redox cycling compounds,[50] this system may play a major role in bacterial resistance to killing by reactive oxygen intermediates generated by phagocytic cells. This may be critical considering that one striking constant of the CF lung is the recurring inflammation and presence of large numbers of neutrophils and macrophages.[17,18]

3. CONTROL OF PILIN AND FLAGELLIN SYNTHESIS

The role of the pili and the single polar flagellum appendages in the pathogenesis of *P. aeruginosa* has been well established. Pili facilitate attachment of this bacterium to eukaryotic cells,[5,53] while flagella are important for invasive virulence of *P. aeruginosa*.[54] Both organelles are required for full virulence in the burned mouse model. *P. aeruginosa* pilin belongs to the type IV pili, which along with *Neisseria gonorrhoeae, Neisseria meningitidis, Vibrio cholerae*, and several other Gram-negative bacteria, are processed via a specific type IV prepilin peptidase (PilD).[55] The corresponding prepilins, although having a characteristic signal sequence, unlike typical bacterial secretory proteins are processed before the signal peptidase site at the N-terminal side of the hydrophobic region and the mature pilins start with an N-methylated Phe or Met residue. As in *Neisseria*, transcription of the pilin gene in *P. aeruginosa* is dependent upon an alternative sigma factor σ^{54} (RpoN)[56] and a two-component signal transduction system PilR-PilS[57,58] (Fig. 3.1B). As with the other systems dependent on response regulators which activate σ^{54}-dependent promoters, in addition to the canonical N-terminal domain, PilR has the central domain containing the ATP-binding motif believed to play a role in open complex formation in the process of transcriptional initiation[46] typical of this subclass of response regulators (RO$_{IV}$).[45] It will be of interest to determine the environmental signals recognized by this system, and how they relate to the pathogenesis of *Pseudomonas* infections.

Flagellin synthesis also depends on σ[54], but this effect is indirect since the flagellin gene promoter has no σ[54] consensus sequence. Instead, the flagellin gene is transcribed by another alternative sigma factor,[59] the *P. aeruginosa* equivalent of σ[28] (FliA) (Fig. 3.1A). No signal transduction elements have been reported to control the flagellin gene in *P. aeruginosa*. However, σ[54] is indirectly involved since mutations in *rpoN*, the gene encoding this sigma factor, affect *fliA* transcription. Because the σ[54] RNA polymerase holoenzyme invariably works in concert with response regulators from the RO$_{IV}$ subfamily, existence of such a regulator has been postulated in this system.[59] The regulation of the flagellin gene by σ[28] is particularly interesting, since it has been shown in *Salmonella typhimurium* that this sigma factor is sequestered by its cognate antisigma factor FlgM[60] thus controlling its activity and flagellin synthesis. Intracellular concentration of FlgM is linked to the morphogenesis processes of the hook flagellin assembly.[60] Control by FlgM also appears to be critical for *Salmonella* virulence.[61] Although little is known about these aspects of the regulation by σ[28] in *P. aeruginosa*, it is likely that an equivalent system exists in this organism. Control by alternative sigma factors and their cognate negative regulators seems to be emerging as yet another recurring theme in the regulation of bacterial virulence. Another example has already been addressed in the section on the control of alginate synthesis and conversion to mucoidy in *P. aeruginosa* (Figs. 3.1E and 3.4).

4. REGULATION OF PHOSPHOLIPASE C BY PHOSPHATE AND OSMOLARITY

Phospholipase C (PLC) plays an important role in bacterial pathogenesis in various contexts of host-parasite interactions. *P. aeruginosa* produces two PLCs: PLC-H, which is hemolytic for human erythrocytes, and PLC-N, which is non-hemolytic.[13] Both PLC-H and PLC-N hydrolyze phosphatydilcholine, a major component of human lung surfactant, releasing phosphorylcholine and diacylglycerol. However, these enzymes have differential activity profiles with other substrates.[13] For example, PLC-H is active with phospholipids containing quaternary ammonium groups such as sphingomyelin, found in the outer leaflet of the erythrocyte membrane. This property is believed to be associated with the hemolytic activity of PLC-H. In contrast, PLC-N is inactive with sphingomyelin. It is however active with phosphatidylserine, found in the inner leaflet of the erythrocyte membrane, which in turn is not a good substrate for PLC-H.

In the case of respiratory tract infections in CF, it has been proposed that PLCs, along with the extracellular alkaline phosphatase, participate in: (i) inorganic phosphate acquisition, and (ii) generation of osmoprotectants such as choline and its derivatives, which may play a role during growth in the dehydrated environment of the CF lung.[62] The combined action of PLC-H, PLC-N, and extracellular alkaline phosphatase, may enhance degradation of eukaryotic membrane phospholipids by sequential action of enzymes. In one scenario,[13] PLC-H, which is more active with the outer leaflet phospholipids, could expose the inner membrane leaflet, allowing degradation of phosphatidylserine by PLC-N. The final product of phospholipid hydrolysis, phosphorylcholine, is then digested by alkaline phosphatase producing: (a) inorganic phosphate, which is used for growth, and (b) choline, which can act as an osmoprotectant or is converted into glycine betaine, the preferred osmoprotectant for *P. aeruginosa*. Although not explored, there could be other effects on eukaryotic cells, e.g. through products of phospholipid hydrolysis such as diacylglycerol or ceramide that may alter eukaryotic signal transduction processes or synergize with inflammatory cytokines such as TNFα.

PLC-H and PLC-N are encoded by the two unlinked genes, *plcS* and *plcN*, respectively.[13] The name for the structural gene encoding PLC-H, *plcS*, reflects the presence of another gene in this operon, *plcR*, proposed to play a role in activating PLC-H. The *plcS* and *plcN* genes, along with the gene encoding extracellular alkaline phosphatase are under regulation by a PhoB-PhoR signal transduction system and are induced in response to phosphate limitation.[62,63] Under phosphate replete conditions (higher than 10 mM), *plcS* is repressed, but nevertheless a significant activity is detectable. In contrast, under such conditions *plcN* is almost completely silent. When inorganic phosphate is below 1 mM, this results in a moderate 5- to 6-fold induction of *plcS*[62,63] while *plcN* undergoes a more dramatic upshift of 50-fold.[62] The PhoB of *P. aeruginosa* is similar to its *E. coli* equivalent,[63] the two are functionally interchangeable in *P. aeruginosa*,[63] and the presence of sequences closely matching the recognition sequence for the *E. coli* PhoB (the *pho* box) has been noted within the promoter regions of *P. aeruginosa phoB*, *plcN*, *plcS* and *oprP*, another phosphate regulated gene.[63] Thus, it is likely that many of the phenomena noted in the *E. coli phoB-phoR* system will be applicable to *P. aeruginosa*, perhaps with some subtle variations such as the location of PhoB binding sites. Also, the observations that osmolarity and osmoprotectants modulate expression of *plcS* and *plcN*[62] suggest that additional regulatory mechanism may cooperate with PhoB-PhoR in control of these genes. The interesting link between phospholipase activity, lung surfactants,

and generation of osmoprotectants, is certainly deserving further analyses in the context of the dehydrated mucus and osmolyte rich environment in the CF lung. As a curiosity, it may also be of interest to mention here that the *phoB* gene from *P. aeruginosa* served to clone its homologue (*mtrA*) from *Mycobacterium tuberculosis* H37Rv, which is the first response regulator characterized in this organism.[64]

5. CONTROL OF OTHER VIRULENCE FACTORS IN *PSEUDOMONAS*: SIGNALING BEYOND TWO-COMPONENT SYSTEMS

Not all virulence factors are under control by the two-component signal transduction systems (Fig. 3.1). As mentioned in the opening paragraph of the Introduction to this chapter, the gene for exotoxin A, *toxA*, is under complex iron regulation which includes the *P. aeruginosa fur* gene as a central factor.[4] The second ADP-ribosylating enzyme (exoenzyme S) produced by *P. aeruginosa* appears to be controlled by an AraC homolog.[65]

Some exoproducts, such as the protease elastase, are under control by an autoinduction system homologous to the LuxR-LuxI family of regulators (Fig. 3.1C).[66] This system gauges the cell density and controls the synthesis of proteases in much the same way as *Vibrio fischeri* controls its luminescence. Both *P. aeruginosa* and *V. fischeri* produce a diffusable substance, called autoinducer, which accumulates in the medium during growth.[67] In both systems the autoinducer, an L-homoserine lactone derivative, is believed to be generated from lipid biosynthetic intermediates and S-adenosylmethionine[67] with participation of the *luxI* (or *lasI*) gene product (the autoinducer synthase). At high cell densities, autoinducer is present in significant concentrations outside and inside the cells. It interacts with LuxR (or LasR), which in turn activates the subordinate genes. *P. aeruginosa* LasR and LasI are 27% and 34.6% identical to LuxR and LuxI from *V. fischeri*, respectively.[66] In addition, the proposed binding sites for LuxR and LasR are relatively well conserved, and can be found centered around -40 relative to the mRNA start sites of the respective genes: *luxI* in *V. fischeri* and *lasB* (the gene encoding elastase) in *P. aeruginosa*.[67] This interesting type of regulation may also affect expression of some other virulence genes in *P. aeruginosa*.

In conclusion, this brief and by no means complete tour of regulatory elements controlling *P. aeruginosa* virulence determinants underscores the diversity of processes participating in the orchestration of their expression in this bacterium. Sometimes this expression is coordinate. In such instances, iron availability, oxidative stress, or osmolarity are more often than other conditions used as environmental cues. However, it is a more frequent occurrence that different virulence factors are independently controlled. This perhaps reflects the diversity of infections that *P. aeruginosa* can cause and the niches that this versatile organism can colonize, requiring maximum flexibility in virulence gene expression.

ACKNOWLEDGMENTS

Research in this laboratory is sponsored by NIH grants AI31139 and AI35217, by Cystic Fibrosis Foundation Grants G229 and G844 and by a grant from Texas Higher Education Coordinating Board.

REFERENCES

1. Iglewski B. Probing *Pseudomonas aeruginosa,* and opportunistic pathogen, ASM News 1989; 55:303-307.
2. Nicas TI, Iglewski B. The contribution of exoproducts to virulence of *Pseudomonas aeruginosa.* Can J Microbiol 1985; 31:387-392.
3. Wick MJ, Frank DW, Storey DG et al. Structure, function and regulation of *Pseudomonas aeruginosa* exotoxin A. Annu Rev Microbiol 1990; 44:335-363.
4. Prince RW, Cox CD, Vasil ML. Coordinate regulation of siderophore and exotoxin A production: molecular cloning and sequencing of the *Pseudomonas aeruginosa* fur gene. J Bacteriol 1993; 175:2589-2598.
5. Woods DE, Strauss DC, Johansom Jr WG et al. Role of pili in adherence of *Pseudomonas aeruginosa* to mammalian buccal epithelial cells. Infect Immun 1980; 29:1146-1151.
6. Simpson DA, Ramphal RR, Lory S. Genetic analysis of *Pseudomonas aeruginosa* adherence: distinct genetic loci control attachment to epithelial cells and mucins. Infect Immun 1992; 60:3771-3779.
7. Galloway D. *Pseudomonas aeruginosa* elastase and elastolysis: recent developments. Mol Microbiol 1991; 5:2315-2321.
8. Toder DS, Ferrell SJ, Nezezon JL et al. *lasA* and *lasB* genes of *P. aeruginosa*: analysis of transcription and gene product activity. Infect Immun 1994; 62:1320-1327.
9. Okuda K, Morihara K, Atsumi Y et al. Complete nucleotide sequence of the structural gene for alkaline proteinase from *Pseudomonas aeruginosa* IFO 3455. Infect Immun 1990; 58:4083-4088.
10. Britigan BE, Rasmussen GT, Cox CD. *Pseudomonas* siderophore pyochelin enhances neutrophil-mediated endothelial cell injury. Am J Physiol 1994; 266:192-198.
11. Britigan BE, Roeder TL, Rasmussen GT et al. Interaction of the *Pseudomonas aeruginosa* secretory product pyocyanin and pyochelin generates hydroxyl radical and causes synergistic damage to endothelial cells. Implications for *Pseudomonas*-associated injury. J Clin Invest 1992; 90:2187-2196.
12. Cryz SJ Jr, Pitt TL, Furer E et al. Role of lipopolysaccharide in virulence of *Pseudomonas aeruginosa.* Infect Immun 1984; 44:508-513.
13. Ostroff RM, Vasil AI, Vasil ML. Molecular characterization of a nonhemolytic and hemolytic phospholipase C from *Pseudomonas aeruginosa.* J Bacteriol 1990; 172:5915-5923.
14. Poole K, Krebes K, McNally C et al. Multiple antibiotic resistance in *Pseudo-*

monas aeruginosa: evidence for involvement of an efflux operon. J Bacteriol 1993; 175:7363-7372.

15. Li X-Z, Livermore DM, Nikaido H. Role of efflux pump(s) in intrinsic resistance of *P. aeruginosa*: resistance to tetracycline, chloramphenicol, and norfloxacin. Antimicrob Agents Chemother 1994; 38: 1732-1741.

16. Collins FS. Cystic fibrosis: molecular biology and therapeutic implications. Science 1992; 256: 774-779.

17. Gilligan PH. Microbiology of airway disease in patients with cystic fibrosis. Clin. Microbiol Rev.1991; 4:35-51.

18. Govan JRW. Alginate biosynthesis and other unusual characteristics associated with the pathogenesis of *Pseudomonas aeruginosa* in cystic fibrosis. In: Griffiths E, Donachie W, Stephen J, eds. Bacterial infections of respiratory and gastrointestinal mucosae. Oxford: IRL Press, 1988:67-96.

19. Gacesa P, Russell NJ. The structure and properties of alginate. In: Gacesa P, Russell NJ, eds. Pseudomonas infections and alginates: biochemistry, genetics and pathology. London: Chapman & Hall, Ltd, 1990:29-49.

20. Simpson JA, Smith SE, Dean RT. Scavenging by alginate of free radicals released by macrophages. Free Radical Biol Med 1989; 6:347-353.

21. Learn DB, Brestel EP, Seetharama S. Hypochlorite scavenging by *Pseudomonas aeruginosa* alginate. Infect Immun 1987; 55:1813-1818.

22. Pier GB, Small GJ, Warren HB. Protection against mucoid *Pseudomonas aeruginosa* in rodent models of endobronchial infections. Science 1990; 249:537-540.

23. Deretic V, Gill JF, Chakrabarty AM. Gene *algD* coding for GDPmannose deydrogenase is transcriptionally activated in mucoid *Pseudomonas aeruginosa*. J Bacteriol 1987; 169:351-358.

24. Franklin MJ, Chitnis CE, Gacesa P et al. *Pseudomonas aeruginosa* AlgG is a polymer level alginate C5-mannuronan epimerase. J Bacteriol 1994; 176:1821-1830.

25. Schiller NL, Monday SR, Boyd CM et al. Characterization of the *Pseudomonas aeruginosa* alginate lyase gene (*algL*): cloning, sequencing, and expression in *Escherichia coli*. J Bacteriol 1993; 175: 4780-4789.

26. Franklin MJ, Ohman DE. Identification of *algF* in the alginate biosynthetic gene cluster of *Pseudomonas aeruginosa* which is required for alginate acetylation. J Bacteriol 1993; 175:5057-5065.

27. Shinabarger D, Berry A, May T et al. Purification and characterization of phosphomannose isomerase-guanosine phospho-D-mannose pyrophosphorylase: a bifunctional enzyme in the alginate biosynthetic pathway of *Pseudomonas aeruginosa*. J Biol Chem 1991; 266:2080-2088.

28. Coyne MJJr, Russell KS, Coyle CL et al. The *Pseudomonas aeruginosa algC* gene encodes phosphoglucomutase, required for the synthesis of a complete lipopolysaccharide core. J Bacteriol 1994; 176:3500-3507.

29. Ye RW, Zielinski NA, Chakrabarty AM. Purification and characterization of phosphomannomutase/phosphoglucomutase from *Pseudomonas aeruginosa* involved in biosynthesis of both alginate and lipopolysaccharide. J Bacteriol 1994; 176:4851-4857.

30. Chitnis CE, Ohman DE. Genetic analysis of the alginate biosynthetic gene cluster of *Pseudomonas aeruginosa* shows evidence for an operonic structure. Mol Microbiol 1993; 8:563-590.

31. Deretic V, Schurr MJ, Boucher JC et al. Conversion of *Pseudomonas aeruginosa* to mucoidy in cystic fibrosis: environmental stress and regulation of bacterial virulence by alternative sigma factors. J Bacteriol 1994; 176:2773-2780.

32. Deretic V, Diksit R, Konyecsni WM et al. The *algR* gene which regulates mucoidy in *Pseudomonas aeruginosa* belongs to a class of environmentally responsive genes. J Bacteriol 1989; 171: 1278-1283.

33. Wozniak DJ, Ohman DE. *Pseudomonas aeruginosa* AlgB, a two-component response regulator of the NtrC family, is

required for *algD* transcription. J Bacteriol 1991; 173:1406-1413.

34. Miller JF, Mekalanos JJ, Falkow S. Coordinate regulation and sensory transduction in the control of bacterial virulence. Science 1989; 243:916-922.

35. Mohr CD, Hibler NS, Deretic V. AlgR, a response regulator controlling mucoidy in *Pseudomonas aeruginosa*, binds to the FUS sites of the *algD* promoter located unusually far upstream from the mRNA start site. J Bacteriol 1991; 173:5136-5143.

36. Stock AM, Mottonen JM, Stock JB et al. Three-dimensional structure of CheY, the response regulator of bacterial chemotaxis. Nature 1989; 337:745-749.

37. Deretic V, Leveau JHJ, Mohr CD et al. In vitro phosphorylation of AlgR, a regulator of mucoidy in *Pseudomonas aeruginosa*, by a histidine protein kinase and effects of small phospho-donor molecules. Mol Microbiol 1992; 6:2761-2767.

38. Deretic V, Konyecsni WM. Control of mucoidy in *Pseudomonas aeruginosa*: transcriptional regulation of *algR* and identification of the second regulatory gene, *algQ*. J Bacteriol 1989; 171:3680-3688.

39. Venturi V, Ottevanger C, Leong J et al. Identification and characterization of a siderophore regulatory gene (*pfrA*) of *Pseudomonas putida* WCS358: homology to the alginate regulatory gene *algQ* of *Pseudomonas aeruginosa*. Mol Microbiol 10:63-73.

40. Mohr CD, Leveau JHJ, Krieg DP et al. AlgR-binding sites within the *algD* promoter make up a set of inverted repeats separated by a large intervening segment of DNA. J Bacteriol 1992; 174:6624-6633.

41. Berry A, DeVault JD, Chakrabarty AM. High osmolarity is a signal for enhanced *algD* transcription in mucoid and nonmucoid *Pseudomonas aeruginosa* strains. J Bacteriol 1989; 171:2312-2317.

42. Mohr CD, Martin DW, Konyecsni WM et al. Role of the Far-upstream sites of the *algD* promoter and the *algR* and *rpoN* genes in environmental modulation of mucoidy in *Pseudomonas aeruginosa*. J

Bacteriol 1990; 172:6576-6580.

43. Mohr CD, Sonsteby SK, Deretic V. The *Pseudomonas aeruginosa* homologs of hemC and *hemD* are linked to the gene encoding the regulator of mucoidy *algR*. Mol Gen Genet 1994; 242:177-184.

44. Goldberg JB, Dahnke T. *Pseudomonas aeruginosa* AlgB, which modulates the expression of alginate, is a member of the NtrC subclass of prokaryotic regulators. Mol Microbiol 1992; 6:59-66.

45. Parkinson JS, Kofoid EC. Communication modules in bacterial signaling proteins. Annu Rev Genet 1992; 26:71-112.

46. Weiss DS, Batut J, Klose KE et al. The phosphorylated form of the enhancer-binding protein NTRC has an ATPase activity that is essential for activation of transcription. Cell 1991; 67:155-167.

47. Martin DW, Holloway BW, Deretic V. Characterization of a locus determining the mucoid status of *Pseudomonas aeruginosa*: AlgU shows sequence similarities with a *Bacillus* sigma factor. J Bacteriol 1993; 175:1153-1164.

48. Martin DW, Schurr MJ, Mudd MH et al. Differentiation of *Pseudomonas aeruginosa* into the alginate-producing form: inactivation of *mucB* causes conversion to mucoidy. Mol Microbiol 1993; 9:495-506.

49. Martin DW, Schurr MJ, Mudd MH et al. Mechanism of conversion to mucoidy in *Pseudomonas aeruginosa* infecting cystic fibrosis patients. Proc Natl Acad Sci USA 1993; 90:8377-8381.

50. Martin DW, Schurr MJ, Yu H et al. Analysis of promoters controlled by the putative sigma factor AlgU regulating conversion to mucoidy in *Pseudomonas aeruginosa*: relationship to σ^E and stress response. J Bacteriol 1994; 176.

51. Erickson JW, Gross CA. Identification of the subunit of *Escherichia coli* RNA polymerase: a second alternative factor involved in high-temperature gene expression. Genes Dev 1989; 3:1462-1471.

52. Johnson K, Charles I, Dougan G et al. The role of stress-response protein in *Salmonella typhimurium* virulence. Mol Microbiol 1991; 5:401-407.

53. Sato H, Okinaga K, Saito H. Role of pili in the pathogenesis of *Pseudomonas aeruginosa* burn infection. Microbiol Immunol 1988; 32:131-139.

54. Drake D, Montie TC. Flagella, motility and invasive virulence of *Pseudomonas aeruginosa*. J Gen Microbiol 1988; 134:43-52.

55. Nunn D, Lory S. Component of the protein-excretion apparatus of *P. aeruginosa* are processed by the type IV prepilin peptidase. Proc Natl Acad. Sci USA 1992; 89:47-51.

262:1277-1280.

61. Schmitt CK, Darnell SC, Tesh VL et al. Mutation of *flgM* attenuates virulence of *Salmonella typhimurium*, and mutation of *fliA* represses the attenuated phenotype. J Bacteriol 1994; 176:368-377.

62. Shortridge VD, Lazdunski A, Vasil ML. Osmoprotectants and phosphate regulate expression of phospholipase C in *Pseudomonas aeruginosa*. Mol Microbiol 1992; 6:863-871.

63. Anba J, Bidaud M, Vasil ML et al. Nucleotide sequence of the *Pseudomonas*

CHAPTER 4

PHOP/PHOQ: REGULATING SALMONELLA ADAPTATION TO HOST MICROENVIRONMENTS

Samuel I. Miller

1. INTRODUCTION

S*almonella* serotypes can colonize and infect an enormous range of vertebrate hosts after orally ingestion. Infection results in a wide spectrum of acute and chronic illnesses; including gastroenteritis, enteric (typhoid) fever, and bacteremia.[1] The consequence of *Salmonellae* ingestion by vertebrates depends on the serotype. Certain *Salmonella* serotypes have very narrow host specificity while others infect a wide variety of animal hosts.[1] Examples of host-specific serotypes include *S. typhi* (man), *S. pullorum* (fowl), and *S. arizonae* (reptiles). Such serotypes, with rare exceptions, are only pathogenic for a specific host. In contrast *S. typhimurium* infects a wide variety of hosts, though disease manifestation varies. In immunocompetent humans an acute self limited gastroenteritis is seen after *S. typhimurium* ingestion, while in mice a systemic illness similar to human enteric fever occurs. In addition host immune status dramatically alters disease outcome. Humans with the acquired immunodeficiency syndrome (AIDS) are highly susceptible to severe systemic infection with *S. typhimurium*[2] and the susceptibility of inbred mice to this serotype varies widely as a result of mutations that alter immune function.[3,4]

An interesting research goal is the determination of the mechanisms involved in *Salmonellae* host specificity. The natural diversity of host-*Salmonellae* interactions suggests that multiple bacterial and host factors contribute to pathogenesis. Important interactions of *Salmonellae* with two host cell types, macrophages and epithelial cells, appears to contribute to this diversity. The ability to survive phagocytosis by macrophages is an essential factor in mouse susceptibility to *S.*

Signal Transduction and Bacterial Virulence, edited by Rino Rappuoli, Vincenzo Scarlato and Beatrice Aricò. © 1995 R.G. Landes Company.

typhimurium and correlates with bacterial pathogenicity.[3,5] Perhaps most convincing for the involvement of macrophages in host specificity is documented by the susceptibility of inbred mice to *S. typhimurium,* which is in part due to a genetic defect in the ability of macrophages to restrict bacterial growth.[3] This defect, which is inherited in an autosomal dominant fashion, is most likely due to a mutation in the *Nramp* gene localized on chromosome 1.[6]

Because of the initial interaction of enteric pathogens with the host at mucosal surfaces, *Salmonellae*-epithelial cell interactions are likely to contribute to host-pathogen specificity. In contrast to macrophages, no data with inbred genetic hosts documents this importance. Despite this limitation, recent results demonstrating an association between the ability of *Salmonella* serotypes to signal epithelial cells and to cause human gastroenteritis support an important role for epithelial cells in serotype human specificity.[7] Remarkably, a single bacterial regulatory network plays a prominent role in modulating *Salmonellae* interactions with both macrophages and epithelial cells. This network, the PhoP regulon, is the subject of this chapter.

The PhoP regulon is defined as genes that encode two regulatory proteins (PhoP/ PhoQ) and the genes these proteins transcriptionally regulate in response to environmental change (Fig. 4.1). This regulon facilitates bacterial adaptation to environments the organism encounters as it infects mammalian tissues, including the harsh environments found within acidified macrophage phagosomes. The PhoP/ PhoQ proteins can activate as well as repress transcription of target genes. Genes whose transcription is activated in the presence of these proteins are termed *pag* for *phoP* activated genes. Those that are transcriptionally repressed under conditions of *pag* activation are termed *prg* for *phoP* repressed genes. Though much remains to be learned about the PhoP regulon and its role in virulence, the study of this regulon has aided the development of a number of in vitro assays of *Salmonellae*

virulence and should lead to further discovery and characterization of bacterial factors that promote *Salmonellae* pathogenesis and host specificity.

2. THE PHOP/PHOQ REGULATORS

PhoP and PhoQ are members of a family of regulatory proteins that, in response to environmental change, activate gene transcription through a phosphotransfer mechanism.[8,9] PhoQ is an environmental sensor and PhoP is a transcriptional activator. The biochemical mechanism of members of this family has been studied in detail and by analogy to these systems on sensing the macrophage intracellular environment PhoQ-phosphate is formed by phosphorylation of a histidine contained within a consensus five amino acid sequence. Then PhoQ transfers phosphate to an aspartate(s) in the amino terminal domain of PhoP. PhoP-phosphate binds to DNA to promote transcription of target genes.

The *phoP* and *phoQ* genes are located in an operon at 25 minutes on the *S. typhimurium* chromosome.[8] The *phoP* locus was originally discovered by Ames and collaborators as one of two loci, the other being *phoN* located at 95 minutes on the chromosome, essential to the production of a periplasmic acid phosphatase.[10] Based on the deduced amino acid sequences the predicted products of the *phoP* and *phoQ* genes are 224 and 487 amino acids in length.[8] Consistent with this data, antisera produced against peptide sequences of PhoP and PhoQ recognize proteins that are approximately 28 and 55 kDa in size (Miller S and Hohmann L, unpublished data).

PhoQ contains two hydrophobic regions, amino acids 17-44 and 191-218, that are predicted to be transmembrane domains.[8] The 147 amino acid hydrophilic segment between these domains is predicted to be located in the bacterial periplasm and the remainder of PhoQ is predicted to be localized in the cytoplasm. Consistent with this predicted PhoQ topology is data from a topologic analysis

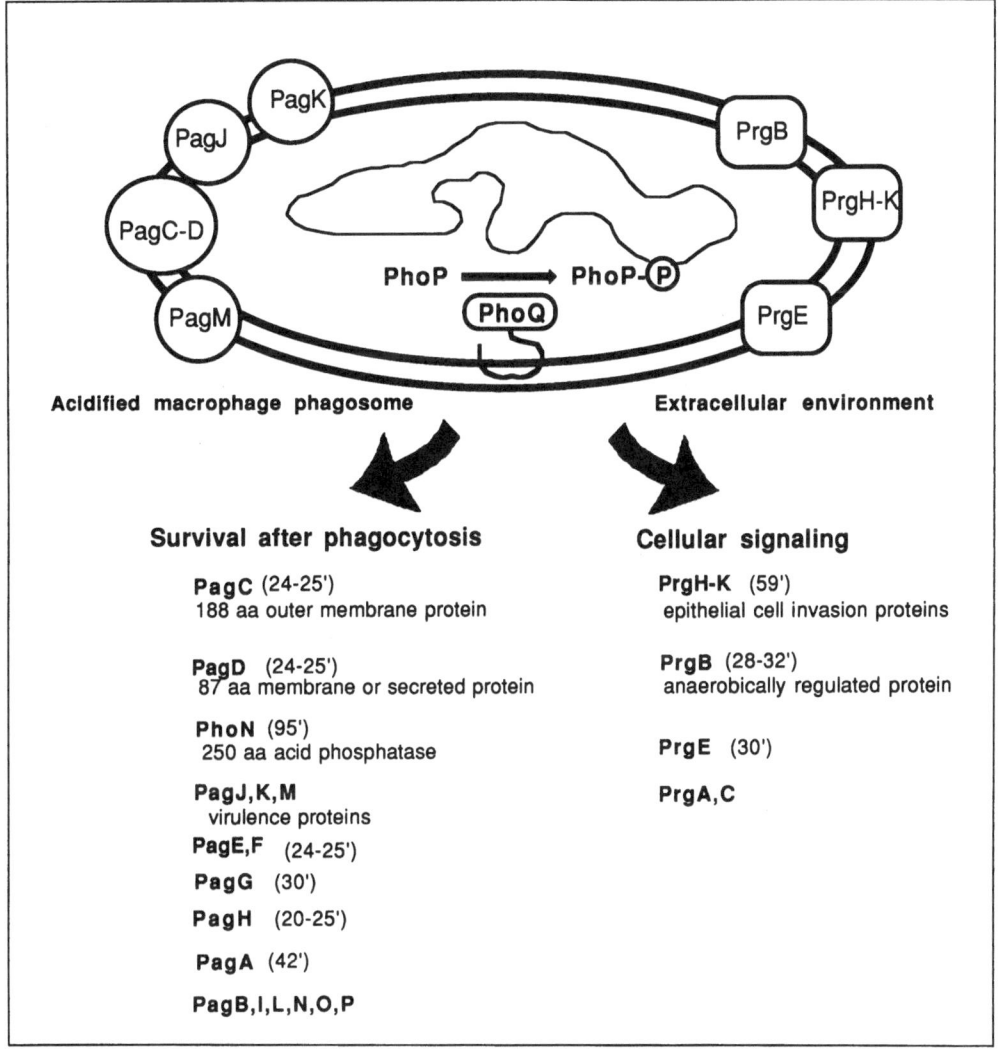

Fig. 4.1. A model representation of the PhoP regulon. PhoQ is a membrane associated sensor-kinase which phosphorylates PhoP to activate and repress target genes within acidified macrophage phagosomes.

of Tn*phoA*-generated alkaline phosphatase fusion proteins. Enzymatically active gene fusions have Tn*phoA* inserted within the periplasmic domain (active alkaline phosphatase activity can only occur when the *phoA* encoded portion of such fusion proteins is exported from the cytoplasm into the periplasm). Consistent with the specificity required of a sensor-kinase, PhoQ has similarity to other histidine-kinases only in this cytoplasmic domain at the predicted site of autophosphorylation and at the carboxy-terminus where ATP binding and phosphatase activity have been localized in other systems.[8]

Since mutations in the PhoQ periplasmic domain can either activate or abolish PhoQ regulated gene expression (Miller S and Hohmann L, unpublished data) this suggests that this domain has a role in environmental signaling, either by direct binding of a signaling molecule or through conformational change. One such allele of the *phoP* locus (*pho-24*) has greatly facilitated the study of the PhoP regulon and its role in bacterial virulence. Strains containing the *pho-24* allele have a PhoP constitutive (PhoP^C) phenotype with resultant increase in *pag* expression and *prg* repression.[11] This phenotype is the result of a single amino acid change that alters a threonine in the periplasmic domain of PhoQ (Miller S and Hohmann L, unpublished data).

Similar to all members of this transcriptional activator family, PhoP has a conserved amino-terminal domain that is predicted to be phosphorylated at an aspartate(s) by PhoQ.[8] The carboxyl-terminus of PhoP is similar to that of a subset of this family of proteins (OmpR subclass) and the amino-terminus of the *Vibrio cholerae* transcriptional activator ToxR.[9] This conserved domain has been demonstrated in a number of systems to be essential to transcriptional activation from promoters that recognize the major form of RNA polymerase corresponding to the sigma 70 of *Escherichia coli*. This domain has also been demonstrated in a number of systems to bind to 5' controlling regions of target genes. Therefore phosphorylation of the amino-terminal domain of PhoP is predicted to alter the conformation of the carboxy-terminal domain to facilitate DNA binding to target genes. Similar to *ompR*, *phoP* transcription is neither autoregulated or induced by environmental signals. Therefore, PhoP copy number remains constant and the amount of PhoP phosphorylation regulates transcription of *pag* and *prg* (Miller S, unpublished data).

3. PhoP/PhoQ REGULATED GENES

Analysis of the proteins altered in synthesis in PhoP[C] mutants predicted that approximately 40 proteins (20 activated and 20 repressed) were regulated by PhoP/PhoQ.[11] To date, 17 *pag* and 5 *prg* loci have been defined that encode at least 23 membrane or secreted proteins.[8,10,12-15]

Ames and colleagues originally defined two loci, *phoP* and *phoN*, as required for the synthesis of a periplasmic acid phosphatase.[10] *phoN* encodes an acid phosphatase of pH optimum 5.5 that has no substrate specificity.[16,17] This enzyme functions as a homodimer of 27 kDa subunits. The DNA sequence of *phoN* has been

determined and the deduced amino acid sequence, consistent with previous work on purified protein, predicts that the acid phosphatase is a 250 amino acid polypeptide with an amino-terminal leader sequence that targets the enzyme for secretion.[18,19] *phoN* has a G+C content of 43%, which is significantly lower than that of the G+C content (51%) of the *S. typhimurium* genome overall. Only two other Gram's stain-negative bacteria with low G+C content, *Morganella morganii* and *Provedencia stuartii*, have been demonstrated to have nonspecific acid phosphatase enzymatic activity. This data, combined with the observation that DNA similar to the *oriT* of IncFII plasmids is located 5' to *phoN*, led Groisman and colleagues to speculate that *phoN* was inherited by lateral transmission from a low G+C organism in a plasmid mediated event.[19]

The acid phosphatase has been thought to function as a scavenger of phosphate groups and esters which are not transportable across the inner membrane. Such an enzyme could provide a survival advantage under phosphate limiting conditions. However, in contrast to what has been observed with the analogous *E. coli* enzyme of different pH optimum, alkaline phosphatase, *phoN* mutants grow more slowly only when unusual phosphomonoesters are provided as the sole phosphate source.[10] Similar to other PhoP-activated genes, *phoN* is predicted to be induced after bacteria are phagocytosed by macrophages and it is possible that this enzyme functions to scavenge phosphates within macrophage phagosomes.

In addition to *phoN*, sixteen distinct *pag* loci have been defined as recorder gene fusions that have abolished or significant reduction in fusion protein production on deletion or null mutation of the *phoP* locus.[8,12] These loci have been arbitrarily designated *pagA-P* based on the order in which they were identified. *pagA* and *pagB* were defined as transcriptional gene fusions to β-galactosidase generated by MudJ mutagenesis, and the remainder of the *pag* loci have been defined as alkaline phosphatase translational gene fusions generated by Tn*phoA* mutagenesis. With the exception of *phoN*, *pagC*, and *pagD*, the *pag* loci have not been extensively characterized at the molecular level.

pag have a diverse chromosomal location though four are clustered in the region of 24-25 minutes. The location of these loci has been defined by P22 bacteriophage linkage analysis using strains AK3233 and AK3140 whose transposon insertions are known to be located at 24-25 minutes.[8,12] *pagC* and *pagD* which are divergently transcribed and separated by an intergenic region of 134 nucleotides[14] are both similarly linked to the markers of AK3233 (80%) and AK3140 (30%). *pagE* and *pagF* exhibit different linkages to these markers. *pagE* is 39% linked to the marker of AK3233 and 99% linked to the marker of AK3140. In contrast *pagF* is only linked to the marker of AK3140 (31%). Other defined linkages of *pag* include *phoN* (95 minutes), *pagA* (42 minutes), *pagG* (30 minutes), and *pagH* (20-25 minutes). Though precise chromosomal locations have not been determined for the other known *pag*, these loci are not linked to markers that define *pag* with known location.

The *pagC* and *pagD* genes are the best characterized of the *pag* loci and Tn*phoA* insertions in either of these genes reduce the ability of bacteria to survive after phagocytosis by macrophages. *pagC* encodes an 188 amino acid outer membrane protein that has similarity to a family of outer membrane proteins of Gram-negative bacteria.[15] These proteins include Ail, a *Yersinia enterocolitica* invasion protein,[20] Lom, encoded by bacteriophage lambda,[21] OmpX, implicated in antibiotic resistance of *Enterobacter cloacae*,[22] and Rsk, a megaplasmid-encoded protein of *S. typhimurium* implicated in complement resistance.[23] The functions of Ail, OmpX, and Rsk have been defined on the basis of multicopy expression of the genes encoding these proteins in *E. coli* or in the case of Rsk, in rough *S. typhimurium*. The enzymatic or structural role of the PagC family remains to

be defined. These proteins have five highly conserved domains and a transmembrane topology has been proposed for these proteins which predicts that the conserved domains are membrane spanning regions and the variable regions are located on the bacterial surface.[22] This model would be consistent with proteins of the PagC family having diverse functions. However, the most conserved sequence of these proteins [NL(V)KYRYE] is hydrophilic and might form salt bonds or interact with other proteins. It seems unlikely that such a charged domain would span the outer membrane.

pagD encodes an 87 amino acid polypeptide with a hydrophobic domain at the amino terminus predicted to target the protein for secretion.[14] The remainder of the protein is hydrophilic, also consistent with a periplasmic or secreted location. The predicted protein product of *pagD* has no similarity to known proteins in the data base. As with *phoN*, *pagC* and *pagD* have a low G+C content suggesting that this region may have been acquired from a mobile genetic element from another enteric organism. The coordinate regulation of genes that are divergently transcribed may indicate that PagD and PagC have an important stoichiometric relationship regulated by PhoP-phosphate binding to the *pagC*-*pagD* intergenic region.

In contrast to *pag* loci, only five *prg* loci have been defined to date.[13] These also have separate chromosomal location. *prgE* is highly linked to *pagH* at 30 minutes and *prgB* is also located in this region (28-32 minutes). *prgH* is located in the region of 59 minutes closely linked to the *hil* locus, which has been defined as important to bacterial mediated endocytosis by epithelial cells. *prgH* is part of an operon that encodes four proteins designated PrgH-K.[52] PrgH is predicted to be a 392 amino acid protein that is not similar to any proteins in the data base. The deduced amino acid sequences of two of the operon encoded proteins, PrgH and PrgK, predicts that they could be membrane lipoproteins. It is not known at this time which of the proteins in the *prgH* operon are essential for bacterial mediated endocytosis and entrance into epithelial cells. Many of the proteins encoded by genes located in this chromosome region have been implicated in stimulating uptake by epithelial cells and are similar to genes of other Gram-negative bacteria that are essential for secretion of virulence factors.[24,25,26,47] Among these are *Shigella* and *Yersinia sp.* virulence proteins encoded by genes (*spa, mxi, yop*) located on their respective plasmids. Other genes in this chromosomal region may be PhoP-repressed including some of these that encode homologues of *Shigella* (Spa and Mxi) and *Yersinia* (Yops) proteins as a recent analysis (Lee C, personal communication, Harvard Medical School, Boston, MA) has demonstrated that transcription of many genes in the 58-60 minute region can be repressed as part of the PhoP[C] phenotype.

4. VIRULENCE PHENOTYPES OF PhoP REGULON MUTATIONS

Both PhoP-null and PhoP[C] mutants of *S. typhimurium* are markedly attenuated for BALB/c mouse virulence as well as survival within cultured macrophages.[8,11,27,28] PhoP null mutations increase the LD_{50} of *S. typhimurium* for mice by greater than 500,000-fold while strains with PhoP[C] mutations have an LD_{50} greater than 10,000 times that of wild type bacteria. The affect of PhoP-null mutations is most likely due to the inability to express *pag*. Consistent with this supposition, strains with transposon insertions in *pagC, pagD, pagJ, pagK,* and *pagM* all have elevated LD_{50} greater than 1000 times that of wild type organisms and are defective in survival within cultured macrophages.[8,12] Transposon insertions in *phoN* and the other defined *pag* do not significantly alter mouse virulence.[8,12,28] This work seems to indicate that a subset of *pag* are important to virulence. However, it is possible that *pag* encode redundant functions or that mouse virulence assays are insensitive for minor reductions in bacterial virulence. Despite the lack of

evidence for involvement of *pagA* and *pagB* in virulence, these genes, and probably all *pag*, are induced after phagocytosis by macrophages.[29] This fact alone seems to indicate a role for these genes in survival within macrophages.

Consistent with the requirements of PhoP-activated genes for survival within macrophages, PhoP and PhoQ mutants are more sensitive to killing by cationic peptide antibiotics[28,30] and are more sensitive to pH < 3.3.[31] Such mouse cationic peptides (cryptdins) have only been detected in the intestinal lumen as a result of Paneth cell secretion.[32] However, it is likely that similar or related peptides may exist in mouse macrophages. Though pH < 3.3 is an environmental condition that would only be encountered by bacteria in the lumen of the stomach, resistance to pH < 5.0 would be important to survival within macrophages after full phagosome acidification.[29]

The virulence defect of PhoP[C] mutants could simply be a result of the inability to expression *prg*. To date *prgH* is the only *prg* defined as a virulence gene,[13] though only a small number of *prg* have been defined. Strains with transposon insertions in *prgH* have a minor virulence defect as measured by mouse LD_{50}, though these mutants have a significant defect in crossing the intestinal mucosal barrier. Consistent with this defined role, *prgH* mutants are defective in signaling epithelial cells to phagocytose bacteria.[13]

Perhaps more than simply lack of expression of *prg*, the ability to quickly express and/or repress PhoP-regulated virulence genes at the proper time in the infectious cycle could be an essential virulence property. Consistent with this possibility PhoP[C] mutants are defective in survival early after phagocytosis by macrophages, while PhoP-null mutants are defective later after phagocytosis, coincident with acidification of the phagosome.[33]

PhoP[C] mutants are sensitive to killing by human serum, though this unlikely plays a direct role in mouse virulence as PhoP[C] mutants are no more sensitive to mouse serum than wild type bacteria (Peguses D, Behlau I, Miller SI, unpublished data). The heat lability of serum killing supports the likelihood that these mutants are more sensitive to complement. Both complement and cationic antimicrobial peptides, such as the defensins, are membrane active agents and defensins have been demonstrated to form anionic selective channels in black lipid bilayers.[34] Rough mutants defective in lipopolysaccharide chain length are more sensitive to both of these agents. This is presumably because of greater access of these membrane active molecules to the outer membrane. Since, both PhoP[C] and PhoP-null bacteria have normal lipopolysaccharide chain length and normal levels of enzymes required for lipopolysaccharide biosynthesis (Miller S and Osborn M, unpublished data), the changes in the bacterial envelope that result in sensitivity to these membrane active molecules are not the result of a major alterations in LPS synthesis. Most likely, this indicates that opposite extremes of PhoP/PhoQ gene regulation cause significant alteration in properties of the bacterial envelope other than LPS. Therefore, this suggests that bacterial survival within macrophages requires the ability to alter the bacterial envelope in response to a changing macrophage phagosome environment. These alterations are most likely the cumulative affect of a number of PhoP-regulated gene products, as no *pag* or *prg* mutants defined to date have sensitivity to defensins or human serum.

5. REGULATION OF PhoP-REGULATED GENES

Though a number of bacterial growth conditions can increase PhoP-activated gene expression, none increase transcription to the level seen after *S. typhimurium* phagocytosis by macrophages. Ames and colleagues defined in detail the regulation of acid phosphatase production by *S. typhimurium* and these experiments most likely reflect the transcription of *phoN*.[10] Acid phosphatase production is increased

as bacterial growth is limited (starvation). In contrast to *E. coli* alkaline phosphatase which is specifically induced by phosphate limitation, acid phosphatase is induced regardless of whether the limited substrate is carbon, nitrogen, sulphur, or phosphate. This induction is independent of catabolite repression mediated through adenylate cyclase.[10] Growth on relatively poor growth substrates such as ethanolamine and succinate also increase PhoP-activated gene transcription which reflects the fact that *pag* expression correlates inversely with bacterial growth rate. Consistent with starvation signaling PhoP activation, *pag* expression is induced on entering stationary phase after growth on rich medium.[13,15]

Several hours after *S. typhimurium* phagocytosis by cultured mouse bone-marrow derived macrophages, *pag* expression is dramatically induced approximately 100-fold.[29] The selectivity of the macrophage phagosome environment for *pag* induction is reflected in the fact that no induction of gene expression is noted after *S. typhimurium* uptake by cultured epithelial cell lines. Experiments by Abshire and Neidhardt have shown that the macrophage phagosome environment results in the selective induction of a subset of gene products induced by other stresses such as heat shock, peroxide, and DNA damage. This suggests that *S. typhimurium* have specific recognition systems for this environment.[35] PhoQ appears to be an important sensor of this environment as PagC appears to be among the most highly induced proteins in two-dimensional protein electrophoretic analysis of bacteria harvested from macrophages.[35] In addition, a number of such macrophage-induced proteins are absent when PhoP-null mutants are similarly tested.[36]

The intracellular activation of *pag* transcription requires phagosome acidification.[29] Transcriptional activation is abolished on addition of weak bases which buffer phagosome pH. Measurement of the pH from individual macrophage phagosomes containing *S. typhimurium* indicates that the eventual and delayed acidification of bacteria containing phagosomes to pH < 5.0 correlates with activation of PhoQ-regulated gene expression.[29] Activation of *pag* transcription by acid shock to pH < 5.0 has been demonstrated, however the magnitude of transcriptional induction is approximately 10-fold less then that observed within acidified macrophage phagosomes.[29] This strongly suggests that additional, undefined signals specific to macrophage phagosomes are also important to PhoQ activation and subsequent bacterial survival after phagocytosis.

In general, conditions which repress PhoP activated gene expression are conditions for maximal *prg* expression.[13] Therefore, maximal *prg* expression occurs during logarithmic bacterial growth in rich medium. There is diversity in the conditions for maximal induction of *prg* expression suggesting that PhoP-phosphate has the capability of repression of promoters activated by diverse signals. For instance, *prgB* is highly induced under anaerobic growth conditions and low pH while *prgH* is maximally expressed at neutral pH during aerobic growth.[13] The lack of induction of *prgH* by microaerophilic growth is in contrast to the expression of *org*,[37] another *hil* locus invasion gene, that is induced by microaerophilic conditions.

One model to explain these results predicts that PhoP-phosphate or a *pag* directly binds to specific *prg* DNA regions to interfere with transcription from diverse promoters. This may be analogous to certain *envZ* mutant alleles that though they act through OmpR have pleiotropic negative regulatory effect on the transcription of genes that are part of other regulons.[38,39] The effect of these mutations can be suppressed by mutations in *rpoA*, the gene that encodes the alpha subunit of RNA polymerase.[40] This suggests that OmpR-phosphate directly binds to repressed promoters to interfere with binding of RNA polymerase. These results combined with the fact that at low levels of PhoP-phosphate (repressing conditions) no *prg* repression occurs, lead to a model that predicts that *prg* binding sites have lower affinity for

PhoP-phosphate than *pag* promoters. Under conditions of excess PhoP-phosphate, binding to *prg* DNA occurs which interferes with transcriptional activation.

6. ROLE OF THE PHOP REGULON IN *SALMONELLAE* SIGNALING TO MAMMALIAN CELLS

In recent years the emergence of in vitro models of bacterial infection using cultured mammalian cells has led to the development of several assays of bacterial virulence. Three model systems will be discussed in which strains containing PhoP regulon mutations have been evaluated. The first of these model systems to be discussed utilizes *Salmonellae* and primary cell cultures of mouse bone marrow-derived macrophages. After *Salmonellae* infection, cultures are assayed for the production of a novel intracellular organelle the spacious phagosome (SP). The second model system utilizes cultured nonpolarized epithelial cell line cultures. The ability of bacteria to enter these cells can be scored by their protection from gentamicin killing. The third models human gastroenteritis by utilizing polarized T84 human intestinal-derived epithelial cells, *Salmonellae*, and human polymorphonuclear leukocytes. The ability of bacteria to elicit neutrophil migration across an intact epithelial monolayer is assayed.

After organisms are bound to macrophage surfaces by receptor-ligand interactions they are typically ingested within a phagosome whose membrane closely opposes the microorganism. In contrast to this typical mechanism, exposure of *S. typhimurium* to macrophages results in the stimulation of generalized macropinocytosis and organisms enter macrophages in large fluid filled organelles termed spacious phagosomes (SP) (Fig. 4.2).[29,33] In these phagosomes the membrane is not adherent to *S. typhimurium*, bacteria can swim freely though they may adhere to the phagosome membrane. Macropinocytosis or literally "big drink" involves the internalization of fluid external to the cell by the fusion of membrane ruffles formed as a consequence of cytoskeletal rearrangements.[41] Macropinosome formation can be stimulated by growth factors and transforming agents and SP are morphologically similar these organelles. However, since they contain a particle (bacteria) and may not be biochemically identical to micropinosomes they have been termed SP.[33] While both SP and macropinosomes form similarly and rapidly fuse with the lysosomal compartment,[33,42] macropinosomes shrink and disappear within 15 minutes. In contrast, SP migrate around the macrophage nucleus persisting in the macrophage cytoplasm for 30 minutes to hours after formation. SP can continue to expand and enlarge by fusion with other SP or macropinosomes for several hours, though some SP are observed to shrink within 30 minutes of formation.[33] If external bacteria are removed with gentamicin treatment

this suppresses the signal for membrane ruffling and SP shrinkage ensues. Most but not all SP (approximately 5% of bacteria) have shrunk by the time PhoP-activated gene expression is induced at 4-6 hours after the addition of gentamicin.[33]

In contrast to the observations with wild type organisms, PhoP^C mutants are markedly defective in the induction of macropinocytosis and SP formation.[33] These mutants enter macrophages through a process morphologically identical to that observed for *Yersinia enterocolitica* opsonized with normal mouse serum. This suggests that *prg* products are essential to the stimulation of macropinocytosis and SP formation.

The persistence of *S. typhimurium* within this novel organelle suggests that SP formation is an important mechanism for *Salmonellae* survival within macrophages. The fact that PhoP^C strains are mark-

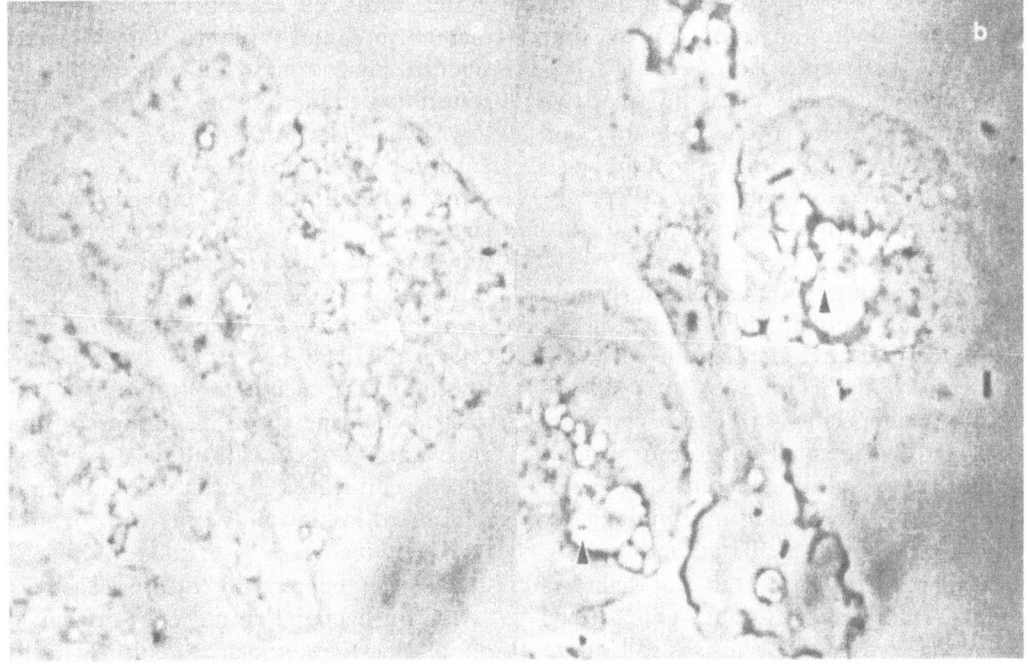

Fig. 4.2. Stimulation of macrophage spacious phagosome (SP) formation, generalized macropinocytosis and membrane ruffling by Salmonella typhimurium. The images are selected frames from time-lapse videomicroscopy before (a) and after (b) exposure of bone marrow-derived macrophages from BALB/c mice to S. typhimurium. In (a) macrophages are seen as relatively quiescent 12 hours after removal from growth factors. In (b) they are shown 25 minutes after exposure to bacteria. The arrows indicate bacteria within SP. The SP and macropinosomes are phase bright organelles that are clustered around the macrophage nuclei. The membrane ruffles are phase dark wavy lines near cell borders.

edly defective in survival within macrophages early after phagocytosis supports the likelihood that the SP is important to pathogenesis.[33] Additionally, the exposure of a variety of *Salmonella* serotypes of varied mouse pathogenicity to macrophages from mouse strains with different *Salmonella* susceptibility demonstrate that host-pathogen specificity roughly correlates with the ability to form SP.[43] These data also suggest that host specificity of *Salmonella* serotypes may in part be a direct result of signaling at the macrophage cell surface to form SP.

An interesting area for speculation is how does the SP promote bacterial survival within phagosomes? SP formation and maintenance could simply dilute toxic lysosomal contents, and volume of SP may explain the delayed acidification of *Salmonella*-containing phagosomes. In addition the SP likely provides a transition environment to allow the organism to acclimate to the phagosome. Finally, the permissive environment of the SP may allow bacterial replication. The observations of Abshire and Neidhardt[35] that two populations of bacteria (one replicating and another static) exist within macrophages may be a reflection that some SP persist and others rapidly shrink within 30 minutes of formation.

Bacterial internalization and formation of SP occurs both in the absence of opsonization and after a variety of opsonization conditions.[33] However, after opsonization with specific antibody bacteria are phagocytosed in small phagosomes whose membrane is tightly adherent to the microorganism in a fashion morphologically similar to receptor-mediated endocytosis. Despite the initial uptake of bacteria within small phagosomes, generalized macrophage macropinocytosis is still stimulated. Over minutes these tight adherent phagosomes enlarge to form SP in part by fusion with other SP or macropinosomes. Therefore normal receptor mediated endocytosis is not repressed by *S. typhimurium*, and regardless of the mechanism of entry, SP formation occurs.

In situations in which SP formation was observed by videomicroscopy with a nonpermissive mouse or *Salmonella* serotype, a phenomena of rapid shrinkage of SP has been observed.[43] These shrinking SP collapse and lose volume within 10 seconds, suggesting that water is being forced from these organelles in a mechanical fashion. This rapid shrinking could be due to cytoskeletal contraction or to a rapid change in membrane potential that results in a dramatic osmotic gradient between the phagosome and the cytoplasm or external surface. Therefore, this observation, as well as the enlargement of phagosomes formed after bacteria are opsonized with antibody, suggest that maintenance of SP may involve *Salmonella* gene products interacting with the phagosome membrane. In summary, three phases of *S. typhimurium* infection of macrophages can be defined by a number of different techniques, SP formation that requires PhoP repressed gene products, SP maintenance, and survival within shrunken acidified phagosomes with subsequent transcriptional induction of *pag* (Fig. 4.3).

Salmonella signaling to a variety of eukaryotic cells results in the stimulation of membrane ruffles and bacterial internalization.[44-47] This is best characterized in studies of *Salmonellae* interaction with epithelium and epithelia derived cell lines. Though *Salmonella* enter most of these cells at relatively low efficiency they enter by stimulating membrane ruffling. In contrast to the generalized membrane ruffling and macropinocytosis noted in macrophages, epithelial cells stimulate ruffling in a localized fashion only in close proximity to *S. typhimurium*.[44-46] The study of this uptake can be trivially measured by assaying for bacterial entrance into normally nonphagocytic nonpolarized epithelial cells. These assays have resulted in the discovery of a number of bacterial genes involved in this process, most of which are similar to a family of genes involved in secretion of virulence factors of *Shigella* and *Yersinia sp*.[24,25,47] PhoP-repressed gene products are essential to this signaling and both PhoP[C]

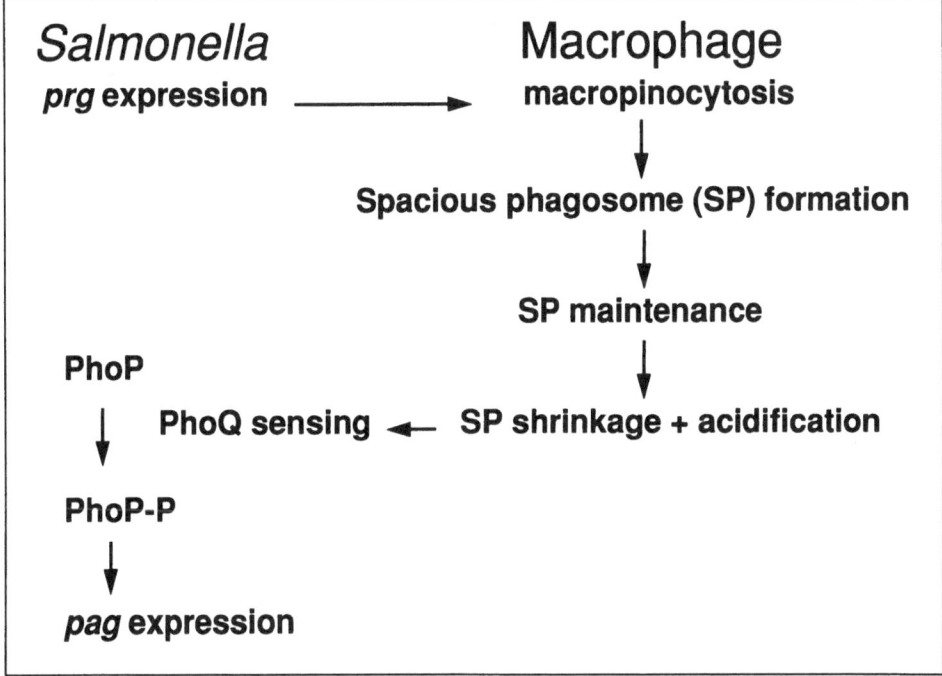

Fig. 4.3. Three phase model of Salmonella-macrophage interactions. (1) PhoP-repressed gene (prg) products facilitate the formation of spacious phagosomes (SP). (2) SP maintenance. (3) SP shrinkage and acidification with activation of PhoQ and subsequent PhoP-activated gene (pag) expression.

mutants as well as *prgH* mutants are defective in this process.[13]

A different or additional set of *prg* products may be required for the stimulation of membrane ruffling by epithelial cells as *prgH* mutants as well as other *Salmonella* mutants defective in stimulation of epithelial cell ruffling and bacterial endocytosis by epithelial cells form equal amounts of SP when compared to wild type organisms by a fluorescence slide assay.[43] This data suggests that perhaps fewer PhoP-repressed gene products are necessary for the induction of macrophage SP formation then for signaling epithelial cells to induce phagocytosis.

The mechanism by which nontyphoidal *Salmonella* serotypes cause diarrhea in humans has long been obscure despite the progress made in understanding the mechanism by which other enteric bacteria, such as *Vibrio cholerae,* cause diarrhea. Recently an in vitro system to model human gastroenteritis has been developed.[48]

Human polarized T84 cells that form tight junctions to create a monolayer with a barrier function have been developed and can be grown on permeable supports with a basolateral and apical compartment. Addition of *Salmonellae* to the apical (lumenal) compartment and human polymorphonuclear leukocytes to the basolateral compartment has allowed the modeling of human nontyphoidal *Salmonella* gastroenteritis in which infiltration of the intestinal mucosa with PMNs is characteristic. After addition of *S. typhimurium* no alteration in chloride secretion can be detected and transepithelial resistance is maintained. However, neutrophil transmigration is stimulated in a physiological direction from the basolateral to apical surface.[48] This migration is independent of bacterial peptide cytokine N-Met-Leu-Phe. Basolateral IL-8 (a powerful cytokine for PMNs) secretion by T84 cells is stimulated, but this gradient is not the signal for transmigration and likely serves to stimulate migration from across the endothelium to the subepithelial area. Therefore a novel signaling to subepithelial neutrophils initiated by nontyphoidal *Salmonella* interacting with epithelial cell apical membranes is responsible for this phenomena.

In contrast, after infection with serotypes such as *S. typhi* and *S. paratyphi* that cause human typhoid fever, a mononuclear cell infiltration localized largely to the Peyer's patches is seen. Interestingly, this model system predicts which *Salmonella* serotypes cause gastroenteritis.[7] In addition *S. typhimurium* PhoP^C and *hil* (*prgH*) mutants as well as other signaling mutants identified by their inability to trigger bacterial mediated endocytosis do not induce neutrophil transmigration. Though all the signaling mutants are defective in entrance into this cell type, serotypes that do not cause human diarrhea and do not cause neutrophil transmigration enter the apical pole of T84 cells equally well to strains that cause diarrhea. The ability to enter these cells from the apical surface seems less important to the pathogenesis of gastroenteritis than correct signaling initiated by the bacteria. Therefore, these results strongly suggest that the ability to elicit transepithelial signaling to neutrophils is essential to the mechanism of human gastroenteritis. In immunocompetent individuals this signaling likely contributes to the development of a self-limited infection. The pathways triggered by *Salmonellae* may utilize similar mechanisms as those responsible for the pathologically similar but non-self-limited inflammation seen in those with Crohn's disease and ulcerative colitis, chronic inflammatory bowel disorders of unknown etiology.

7. UTILITY OF THE PHOP REGULON IN *SALMONELLA* VACCINES

Though both parenteral and live oral vaccines are clinically available for typhoid fever none of these vaccines are widely used as public health tools because of problems related to side effects and efficacy.[1] Therefore, many investigators are

working to develop improved live oral vaccines for both human and animal typhoid fever. Many of the newer human vaccines in development have not seemed to have the proper balance between attenuation and immunogenecity.[49] PhoP regulon mutations could be useful in such vaccines by conferring increased attenuation and safety. In mice both PhoP^C and PhoP mutants have been useful as vaccines.[50] A number of PhoP regulon genes are conserved in *Salmonella typhi* including *phoP*, *phoQ*, and *pagC*. Currently a candidate vaccine deleted for *phoP/phoQ* as well as *aroA*, a biosynthetic gene required for the synthesis of aromatic amino acids, dihydroxybenzoate, and para-aminobenzoate, is being tested in humans (Hohmann L and Miller S, unpublished data). The early data with this live vaccine demonstrate that deletion of *phoP/phoQ* attenuates *Salmonella typhi* indicating that the PhoP regulon functions to promote human virulence. Therefore it is likely that PhoP regulon mutations may have utility in achieving optimal vaccine attenuation and immunogenecity.

Attenuated *Salmonellae* are also attractive as carriers of heterologous antigens. Such *Salmonella*-vectored vaccines may be useful in the development of live oral vaccines that protect against a number of diseases or as boosters to augment previous vaccination. Advantages of the use of *Salmonellae* as an antigen carrier include the ease with which the organism can be genetically manipulated and its large capacity, compared to some virions, for foreign DNA insertion within the bacterial chromosome. In addition, *Salmonellae* have a relative measure of safety for immunosuppressed populations when compared with certain viral and mycobacterial vectors. The development of such systems has been limited by the ineffectiveness of antigen expression from single chromosomal gene copies. Recently this problem has been circumvented by the use of PhoP-regulated promoters.[51] The expression of a model antigen, alkaline phosphatase, from PhoP-regulated promoters has been

Fig. 4.4A

Fig. 4.4B

Fig. 4.4. Cartoon representation of pathogen-host cellular interactions. In (A) a cold war version of these interactions that involves threats of physical violence is shown. In (B) a post cold war modern version that involves subtle signaling between these competitors.

demonstrated to be highly immunogenic and superior to other constitutively expressed promoters in model *Salmonella* vectored vaccines tested in mice. This suggests that the expression of antigens within antigen presenting macrophages can markedly augment the immune response.

8. SUMMARY AND FUTURE DIRECTIONS

The continued characterization of the PhoP regulon should lead to increased knowledge of the molecular basis of intracellular bacterial pathogenesis and the mechanisms by which facultative intracellular bacteria can adapt to reside and replicate within host cells. The current status of our knowledge of *Salmonellae* pathogenesis leads to models in which rather than perceiving bacterial virulence as warfare in which toxin-armed bacteria attack host cells (Fig. 4.4) bacterial virulence can be depicted as subtle signaling between bacteria and host similar to the current state of communication by post cold war economic competitors. Recent data indicate that further work on the signaling between bacteria and their mammalian hosts should increase our knowledge of the specificity of *Salmonella* serotypes for certain hosts. A particularly fertile area for expansion of our understanding of the role such signaling plays in the pathogenic process is its effects on antigen processing and the development of specific immune responses. This knowledge may be useful in the development of live vaccines for salmonellosis and in the development of *Salmonella* as multivalent vectored vaccines for a wide variety of infectious diseases. In addition, since much of the pathogenic process involves tissue damage by stimulation of immune cells such as macrophages and polymorphonuclear leukocytes understanding how bacteria stimulate these processes may lead to new therapies for a variety of autoimmune diseases that currently have no known etiologic cause.

ACKNOWLEDGMENTS

The author would like to thank the Drs. Victor DiRita, Cathy Lee, John Gunn, Libby Hohmann, and Micheal Hantman for helpful comments on this manuscript prior to publication. Particular thanks are directed to Joel Swanson for assistance with the videomicroscopy figure (Fig. 4.2) and to my wife Christine Joly de Lotbiniere for her cartoon drawings (Fig. 4.4). The author is supported by grants AI36516, AI30479, and AI34504 from the National Institutes of Health.

REFERENCES

1. Miller SI, Hohmann EL, Pegues DA. Salmonella (including Salmonella thyphi). In: Mandell GL, Bennett JE, Dolin R, eds. Principles and Practice of Infectious Diseases, 4th ed. 1994:2013-2032.
2. Sperber SJ, Schleupner CJ. *Salmonellosis* during infection with the human immunodeficiency virus. Rev Infect Dis 1987; 9:925-934.
3. Lissner CR, Swanson R, O'Brien A. Genetic control of the innate resistance of mice to *Salmonella typhimurium*: Expression of the Ity gene in peritoneal and splenic macrophages isolated in vitro. J Immunol 1983; 131:3006-3013.
4. Hormaeche CE, Harrington KA, Joysey HS. Natural resistance to *Salmonellae* in mice: control by genes within the major histocompatibility complex. J Infect Dis 1985; 152:1050-1056.
5. Fields PI, Swanson RV, Haidaris CG et al. Mutants of *Salmonella typhimurium* that cannot survive within the macrophage are avirulent. Proc Natl Acad Sci USA 1986; 83:5189-5193.
6. Vidal SM, Malo D, Vogan K, Skamene E, Gros P. Natural resistance to infection with intracellular parasites: Isolation of a candidate for Bcg. Cell 1993; 73:469-485.
7. McCormack BA, Miller SI, Delp-Archer C et al. Transepithelial signalling to neutrophils by *Salmonella*: a novel virulence

mechanism for gastroenteritis. Infect Immun 1995; in press.

8. Miller SI, Kukral AM, Mekalanos JJ. A two-component regulatory system (*phoP/phoQ*) controls *Salmonella typhimurium* virulence. Proc Natl Acad Sci USA 1989; 86:5054-5058.

9. Stock JB, Ninfa AJ, Stock AM. Protein phosphorylation and regulations of adaptive responses in bacteria. Microbiol Rev 1989; 53 :450-490.

10. Kier LD, Weppleman RM, Ames BN. Regulation of nonspecific acid phosphatase in *Salmonella: phonN* and *phoP* genes. J Bacteriol 1979; 138:155-161.

11. Miller SI, Mekalanos JJ. Constitutive expression of the PhoP regulon attenuates *Salmonella* virulence and survival within macrophages. J Bacteriol 1990; 172:2485-2490.

12. Belden WJ, Miller SI. Further characterization of the PhoP regulon: Identification of new *phoP*-activated virulence loci. Infect Immun 1994; 62:5095-5101.

13. Behlau I, Miller SI. A PhoP-repressed gene promotes *Salmonella typhimurium* invasion of epithelial cells. J Bacteriol 1993; 175:4475-4484.

14. Gunn JS, Alpuche-Aranda CM, Loomis WP et al. A virulence gene cluster required for *Salmonella typhimurium* survival within macrophage phasosomes. 1995; submitted for publication.

15. Pulkkinen WS, Miller SI. A *Salmonella typhimurium* virulence protein is similar to a *Yersinia enterocolitica* and a bacteriophage lambda outer membrane protein. J Bacteriol 1991; 173:86-93.

16. Kier LR, Weppelman R, Ames BN. Resolution and purification of three periplasmic phosphatases of *Salmonella typhimurium*. J Bacteriol 1977; 130:399-410.

17. Weppelman R, Kier LD, Ames BN. Properties of two phosphates and a cyclic phosphodiesterase of *Salmonella typhimurium*. J Bacteriol 1977; 130:411-419.

18. Kasahara M, Nakata A, Shinagawa H.

Molecular analysis of the *Salmonella typhimurium phoN* gene, which encodes nonspecific acid phosphatase. J Bacteriol 1991; 173:6760-6765.

19. Groisman EA, Saier MH, Ochman H. Horizontal transfer of a phosphatase gene as evidence for mosaic structure of the *Salmonella* chromosome. EMBO J 1992; 11:1309-1316.

20. Miller VL, Bliska JB, Falkow S. Nucleotide sequence of the *Yersinia enterocolitica ail* gene and characterization of the Ail protein product. J Bacteriol 1990; 172:1062-1069.

21. Barondess JJ, Beckwith J. A bacterial virulence determinant encoded by lysogenic coliphage lambda. Nature (London) 1990; 346:871-872.

22. Stoorvogel, J, van Brussel MJAWM, Tommassen J et al. Molecular characterization of an Enterobacter cloacae outer membrane protein (OmpX). J Bacteriol 1991; 173:156-160.

23. Heffernan EJ, Hardwood J, Fierer J, Guiney D. The *Samonella typhimurium* virulence plasmid complement resistance gene rck is homologous to a family of virulence-related outer membrane protein genes, including *pagC* and *ail*. J Bacteriol 1992; 174:84-91.

24. Galan JE, Ginocchio C, Costeas P. Molecular and functional characterization of the Salmonella invasion gene *invA*: Homology of InvA to members of a new protein family. J Bacteriol 1992; 174-4338-4349.

25. Groisman EA, Ochman H. Cognate gene clusters govern invasion of host epithelial cells by *Salmonella typhimurium* and *Shigella flexneri*. EMBO J 1993; 12:3779-3787.

26. Eichelburg K, Ginocchio CC, Galan JE. Molecular and functional characterization of the *Salmonella typhimurium* invasion genes *invB* and *invC*: Homology of InvC to F_0F_1 ATPase family of proteins. J Bacteriol 1994; 176:4501-4510.

27. Galan JE, Curtiss III R. Virulence and vaccine potential of *phoP* mutants of *Sal-*

monella typhimurium. Microb Pathog 1989; 6:422-443.

28. Fields PI, Groisman EA, Heffron F. A *Salmonella* locus that controls resistance to microbicidal proteins from phagocytic cells. Science 1989; 243:1059-1062.

29. Alpuche Aranda C, Swanson JA, Loomis WP et al. *Salmonella typhimurium* activates virulence gene transcription within acidified macrophage phagosomes. Proc Natl Acad Sci USA 1992; 89:10079-10083.

30. Miller SI, Pulkkinen WS, Selsted ME et al. Characterization of defensin resistance phenotypes associated with mutations in the *phoP* virulence regulon of *Salmonella typhimurium.* Infect Immun 1990; 58:3 706-3710.

31. Foster JW, Hall HK. Adaptive acidification tolerance response of *Salmonella typhimurium.* J Bacteriol 1990; 172:771-778.

32. Selsted ME, Miller SI, Henschen AH et al. Enteric defensins: antibiotic peptide components of intestinal host defense. J Cell Biol 1992; 118:929-936.

33. Alpuche Aranda CM, Racoosin EL, Swanson JA et al. *Salmonella* enter macrophages by macropinocytosis and survive within spacious phagosomes. J Exp Med 1994; 179:601-608.

34. Kagan BL, Selsted ME, Ganz T et al. Antimicrobial peptides form voltage-dependent ion-permeable channels in planar lipid bilayer membranes. Proc Natl Acad Sci USA 1989; 87:210-214.

35. Abshire SK, Neidhardt EC. Growth rate paradox of *Salmonella typhimurium* within host macrophages. J Bacteriol 1993; 175:3744-3748.

36. Buchmeier NA, Heffron F. Induction of Salmonella stress proteins upon infection of macrophages. Science 1990; 248:730-732.

37. Jones BD, Falkow S. Identification and characterization of a *Salmonella typhimurium* oxygen-regulated gene required for bacterial internalization. Infect Immun 1994; 62:3745-3752.

38. Case CC, Bukau B, Granett S et al. Contrasting mechanisms of *envZ* control

of *mal* and *pho* regulon genes in *Escherichia coli.* J Bacteriol 1986; 166:706-7121.

39. Slauch JM, Garrett S, Jackson DE et al. EnvZ functions through OmpR to control porin gene expression in *Escherichia coli* K-12. J Bacteriol 1988; 170:439-441.

40. Garrett S, Silhary TJ. Isolation of mutations in the alpha operon of *Escherichia coli* that suppress the transcriptional defect conferred by a mutation in the porin regulatory gene *envZ.* J Bacteriol 1987; 169:1379-1385.

41. Racoosin EL, Swanson JA. M-CSF-induced macropinocytosis increases solute endocytosis but receptor mediated endocytosis. J Cell Sci 1992; 102:867-880.

42. Racoosin EL, Swanson JA. Macropinosome maturation and fusion with tubular lysosomes in macrophages. J Cell Biol 1993; 121:1011-1020.

43. Miller SI, Alpuche-Aranda C, Berthiaume E, Mock B et al. Spacious phagosome formation correlates with *Salmonella* serotype mouse virulence and inbred mouse susceptibility to *Salmonella typhimurium.* 1995; submitted for publication.

44. Takeuchi A. Electron microscopic studies of experimental *Salmonella* infection I. Penetration into the intestinal epithelium by *Salmonella typhimurium.* Am J Pathol 1967; 50:109-136.

45. Francis CL, Starnbach MN, Falkow S. Morphological and cytoskeletal changes in epithelial cells occur immediately upon interaction of *S. typhimurium* grown under low oxygen conditions. Mol Microbiol 1992; 6:3077-3087.

46. Finlay BB, Ruschkowiscki S, Dedhar S. Cytoskeletal rearrangements accompanying *Salmonella* entry into epithelial cells. J Cell Sci 1991; 99:283-296.

47. Kaniga K, Bossio JC, Galan JE. The *Salmonella typhimurium* invasion genes *invF* and *invG* encode homologues of the AraC and PulD family of proteins. Mol Microbiol 1994; 13:555-568.

48. McCormack BA, Colgan SP, Delp-Archer C et al. *Salmonella typhimurium* attachment to human intestinal epithelial monolayers: transcellular signalling to human subepithelial neutrophils. J Cell

Biol 1993; 123:895-907.

49. Tacket CO, Hone DM, Curtiss III R et al. Comparison of the safety and immunogenecity of Δ*aroC* Δ*aroD* and Δ*cya* Δ*crp* *Salmonella typhi* strains in adult volunteers. Infect Immun 1992; 60:536-541.

50. Miller SI, Loomis WP, Alpuche-Aranda C et al. The PhoP virulence regulon and live oral *Salmonella* vaccines. Vaccine 1993; 11:122-125.

51. Hohmann EL, Oletta CA, Loomis WP et al. Macrophage-inducible expression of a model antigen in *Salmonella typhimurium* enhances immunogenecity. Proc Natl Acad Sci USA 1995; in press.

52. Peguses D, Hautman M, Behlau I, Miller SI. PhoP/PhoQ transcriptional repression of *S. typhimurium* invasion: evidence for a role in protein secretion. Mol Microbiol 1995; in press.

CHAPTER 5

REGULATION OF VIRULENCE IN *VIBRIO CHOLERAE* BY THE TOXR SYSTEM

Victor J. DiRita

1. INTRODUCTION

The bacterium *Vibrio cholerae* causes the human diarrheal disease called cholera. The primary virulence determinant in this pathogen is the cholera toxin, and studies of toxin biogenesis as well as its effects on eukaryotic cells have contributed a great deal to our understanding of fundamental questions in molecular and cellular biology. In addition, *V. cholerae* has become a model organism for studying mucosal pathogenesis and immunity. Work originally aimed at developing toxoid or live oral vaccine preparations to protect susceptible populations against cholera has led to the general question of how toxin and other virulence factors are regulated in *V. cholerae*, which is the subject of this chapter.

Briefly, the regulatory system controlling toxin and colonization factor production that will be reviewed here is controlled by the ToxR and ToxT proteins, which regulate transcription of virulence factor genes in *V. cholerae* (Fig. 5.1). Besides the ToxR/ToxT system, at least two other regulatory systems control aspects of *V. cholerae* virulence. Both of these respond to available iron; one is the Fur/IrgB system leading to expression of an outer membrane protein called IrgA,[1] and the other, required for expression of a hemolysin in some strains of *V. cholerae*, is controlled by the HlyU protein.[2] Neither of these will be discussed in this review.

1.1. OVERVIEW OF TOXINOGENESIS IN *V. CHOLERAE*

Virulent strains of *V. cholerae* cause disease primarily by producing the cholera toxin, a heterooligomer made up of two subunits,

Signal Transduction and Bacterial Virulence, edited by Rino Rappuoli, Vincenzo Scarlato and Beatrice Aricò. © 1995 R.G. Landes Company.

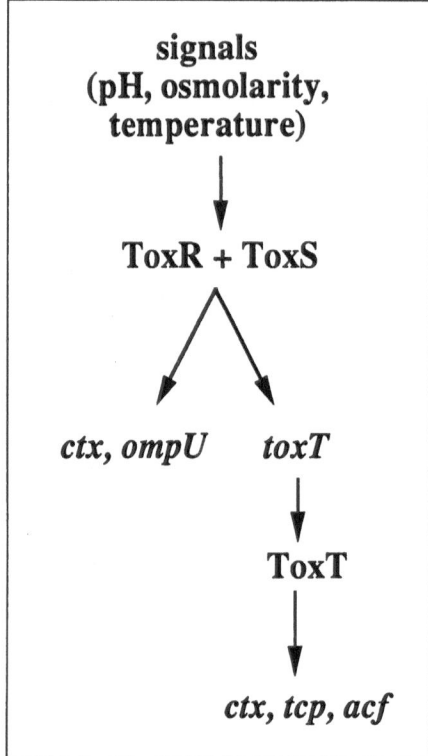

**signals
(pH, osmolarity,
temperature)**

↓

ToxR + ToxS

ctx, ompU   *toxT*

↓

ToxT

↓

ctx, tcp, acf

Fig. 5.1. The ToxR/ToxT regulatory pathway in Vibrio cholerae. *Genes are designated with lower case and italics, proteins with nonitalicized capital letters. The details of the regulatory cascade are explained in the text.*

A and B, present in the ratio of one A to five B subunits.[3] The B subunit of cholera toxin is required for binding to the cellular receptor, ganglioside GM_1, on target intestinal cells, and the A subunit contains the enzymatic portion of the toxin.[4] After cholera toxin binds to target cells and the A subunit is internalized, a proteolytically generated fragment of the A subunit (A1) ADP-ribosylates a specific G protein (Gsα) which results in constitutive adenylate cyclase activity and consequent cyclic AMP (cAMP) elevation.[5] Elevated cAMP disrupts ion transport in intestinal cells resulting in chloride secretion with associated water secretion into the lumen of the intestine,[6] and the overall result for the patient is watery or secretory diarrhea.

Administration of purified cholera toxin to experimental animals results in

symptoms that mimic the disease itself, suggesting that cholera toxin is of primary importance in the pathophysiology of *V. cholerae* infection. Nevertheless, *V. cholerae* produces other determinants of virulence and, with a few exceptions, the virulence factors identified to date in *V. cholerae* are expressed under growth conditions that favor expression of cholera toxin. These culture conditions include growth in rich media (including ~65 mM NaCl or equivalent concentrations of another osmolyte) at 30°C and with a starting pH of 6.5.[7,8] High cholera toxin production in minimal media is achieved when the medium is supplemented with the amino acids glutamate, arginine, serine, and asparagine. Whether any of the in vitro growth conditions that favor high toxin expression reflect the nature of a specific in vivo environment is not known. As will be elaborated on in the following sections, coordinate expression of virulence factors is controlled by the ToxR/ToxT transcription factors (Fig. 5.1).

2. COLONIZATION FACTORS COREGULATED WITH CHOLERA TOXIN IN *V. CHOLERAE*

Unlike many other pathogens reviewed in this book, pathogenesis by *V. cholerae* is restricted to its ability to colonize the host intestinal mucosa and produce toxin. There is no invasion or dissemination of the organism beyond this site. Thus it should not be surprising that virulence factors coregulated with cholera toxin by ToxR and ToxT are colonization factors that enable the organism to establish its niche (Table 5.1). Two major such factors are the toxin coregulated pilus (TCP) and the accessory colonization factor (ACF), both of which are encoded by several genes that were identified by a combination of DNA sequence analysis and by isolating Tn*phoA* mutants in which alkaline phosphatase activity from the fusion protein was expressed under the in vitro growth conditions that favor toxin production.[9-12] In addition to these factors, a major outer

Table 5.1. Coordinately regulated virulence genes controlled by the ToxR/ToxT system

Gene designation	Gene products; functions	References
ctxAB	cholera toxin; responsible for secretory response of intoxicated intestinal cells	4,5,6
tcpA-F, HIJ, PQRST	toxin coregulated pilus subunit (tcpA) and assembly and regulatory factors for Tcp production; colonization factor	9,10,12,15,16,17
tcpG(dsbA)	periplasmic thiol:disulfide bond interchange protein; likely involvement in folding pathway for Tcp, Ctx and protease (not linked to the tcp cluster)	18,19
acfA-D	accessory colonization factors; motility regulators?	11,20
ompU	major outer membrane protein; adhesin?	13,28
tagA	gene product and role in virulence unknown	11
aldA	aldehyde dehydrogenase; role in virulence not known	51

membrane protein, OmpU, is coregulated with toxin and recent characterization of this product suggests that it plays a role in adhesion of *V. cholerae* to eukaryotic cells.[13]

2.1. TCP

The critical role played by TCP in cholera pathogenesis was conclusively demonstrated by Taylor, Mekalanos and colleagues who tested the effects on human volunteers who had ingested mutant strains of *V. cholerae* unable to produce the pilus.[14] The volunteers exhibited few or no symptoms and produced very low titers of detectable antivibrio activity in their sera, observations that suggest TCP is a major factor for efficient colonization by *V. cholerae* and that its expression is a critical requirement of a live, oral vaccine preparation.[14]

According to DNA sequence analysis, 15 open reading frames (ORFs) make up the *tcp* cluster, encoded on a portion of the *V. cholerae* genome of greater than 14 kb in size.[15] These include the gene encoding the major pilus subunit (*tcpA*) as well as those encoding proteins that either have been shown to be or are predicted to be necessary for biogenesis and assembly of the pilus.[15-17] One *tcp* gene identified by Taylor and colleagues, *tcpG*, is not encoded as part of the *tcp* cluster.[18] TcpG has disulfide bond-forming activity and is involved in the folding of both TcpA and Ctx.[19]

Besides the ToxT protein, which is encoded in the *tcp* cluster and which will be discussed in detail below, there are three *tcp* gene products that are postulated to play a role in regulation of Tcp synthesis.[10,15] These are TcpH, TcpP, and TcpI, but of these only TcpI has been characterized to any great extent.[20] This work led to a model for how the signal transduction apparatus governing Tcp mediated colonization and chemotaxis may be directly connected.

TcpI has a domain with strong similarity to the highly conserved domain (HCD) of the Tsr methyl accepting chemotaxis protein of *E. coli*.[20] Mutations in *tcpI* result in increased synthesis of TcpA, the major subunit of the TCP, and also in increased swarming on motility media.[20] That TcpI mutants swarm over a broader area in vitro implies that TcpI functions in the signal transduction mechanism controlling motility and chemotaxis. Peterson and his colleagues proposed a model in which, under appropriate in vivo conditions, there is a switch from motility to colonization and the link between these two properties is TcpI.[20] In the model, TcpI might effect downregulation of motility perhaps by titrating the methylation signal away from the chemotaxis system. The apparent repression of TcpA by TcpI might then be relieved once the cells have stopped swimming and are ready to colonize the epithelial surface. Although this ingenious model suffers from the fact that *tcpI* mutants of *V. cholerae* colonize in the infant mouse with wild type efficiency,[11] it has been established that this animal model may not be the best for studying the role of chemotaxis in infection by *V. cholerae*.[21,22]

2.2. ACF

The *acf* genes are encoded downstream of the *tcp* cluster on the *V. cholerae* genome and the precise role for their products in pathogenesis has not been established. *acf* mutants of *V. cholerae* are not affected in their ability to assemble the TCP, but are attenuated for colonization in the infant mouse model.[13] One of the *acf* genes, *acfB*, encodes a product that, like TcpI, shows similarity to the HCD of methyl accepting chemotaxis proteins (Peterson K, personal communication). Like *tcpI* mutants, cells lacking AcfB also

show increased swarm plate activity but effects on TcpA production have not been reported. Whether TcpI and AcfB have redundant roles in controlling the chemotaxis of *V. cholerae* during infection or whether their similar mutant phenotypes belie distinct in vivo roles has yet to be shown.

2.3. OmpU

The N-terminus of the 38 kDa OmpU protein is reported to show similarity with adhesins from *Bordetella pertussis* and *Haemophilus influenzae*.[13] Additionally, OmpU binds to fibronectin, and antibodies directed against it inhibit adhesion of *V. cholerae* to several eukaryotic cell lines.[13] Taken together, these data suggest that OmpU is an adhesin of *V. cholerae*, but extensive characterization of its role in *V. cholerae* pathogenesis has not been carried out to date as the gene has only recently been cloned.

3. REGULATION OF CTX EXPRESSION BY GENE AMPLIFICATION

The *ctxAB* operon is encoded on an element, called core element, which in many strains of *V. cholerae* is flanked by direct repeats that have features of insertion sequences.[23] In these strains, amplification of the core element occurs during growth of *V. cholerae* in animals during experimental infection, resulting in elevated levels of toxin production.[23] Whether this occurs because of in vivo selection pressures for toxin production or for increased expression of another factor encoded on the core element is not known. At least three other factors thought to be associated with the pathophysiology of *V. cholerae* infection are present on the DNA between the repeated elements. These are *zot* (zonula occludens toxin[24]), *ace* (accessory cholera enterotoxin[25]) and *cep* (core-encoded pilin[26]). While these genes are encoded on the element harboring the *ctxAB* operon, they have not been shown to be part of the coordinate regulatory system that controls cholera toxin.

4. THE ToxR TRANSMEMBRANE TRANSCRIPTIONAL ACTIVATOR

As is the case in many other virulent bacteria, several of which are reviewed in this book, several of the virulence factors of *V. cholerae* are controlled at the level of transcription by a single regulatory gene product, that of the *toxR* gene, which encodes a 32 kDa transcriptional activator protein. ToxR is required in *V. cholerae* for expression of cholera toxin,[27] toxin coregulated pilus,[9] accessory colonization factor,[11] and OmpU.[28] Although transcription of the genes encoding these factors requires ToxR, much of this control is indirect and is the result of ToxR regulation of *toxT* expression (Fig. 5.1; see below).

ToxR was one of the first virulence regulators identified in a bacterial pathogen and is one of the more unusual regulatory proteins described in prokaryotes to date. Unlike many other prokaryotic transcriptional activators, ToxR is an inner membrane that has an amino terminal DNA binding domain that specifically interacts with DNA from the *ctxAB* and *toxT* promoters.[29-31] Its carboxy-terminus is localized to the periplasm by a sequence of hydrophobic amino acids that follow the DNA binding domain. Roughly two-thirds of the protein is cytoplasmic and one third periplasmic, with the transmembrane domain predicted to be 19 amino acids in length.[29]

The DNA binding domain shares similarity to the DNA binding domain of OmpR, the regulator of membrane protein gene expression in *E. coli*.[29,30] Other than this domain, ToxR shares no similarity with members of the sensor-kinase/response regulator family of regulatory proteins described in many bacteria and which are the subject of other chapters in this book. What this implies is that the phosphorylation that controls the activity of OmpR and other such proteins in these two component regulatory systems is not a feature of ToxR activity. The unusual topology for this transcriptional

activator is seen in the CadC protein of *E. coli*, which activates transcription of the lysine decarboxylase operon in response to low pH.[32] To date no other bacterial transcriptional activators have been shown to have the transmembrane topology of ToxR and CadC.

5. PERIPLASMIC INTERACTION BETWEEN ToxR AND ToxS

When the carboxyl terminus of ToxR was replaced with alkaline phosphatase, the resulting fusion protein expressed both ToxR activity and alkaline phosphatase activity.[29,33] This fusion protein also bound specifically to the cholera toxin promoter in band shift experiments.[29] These results suggest that the periplasmic domain of ToxR is dispensable, or that it is insensitive to gross alteration. A clue to the role of this domain in ToxR function has come from analysis of the *toxS* gene product, which is essential for full ToxR activity.

ToxS is predicted to be a 19 kDa, predominantly periplasmic protein with a short cytoplasmic domain followed by a highly hydrophobic stretch predicted to be a membrane localizing sequence.[33] This sequence does not have a consensus signal peptidase cleavage site, suggesting that ToxS is anchored to the membrane, a conclusion that was supported by fractionation studies of ToxS-PhoA fusion proteins.[33] Based on these observations, the most likely place for ToxR and ToxS to interact is in the periplasm, and the result of this interaction is a form of ToxR capable of transcription activation.

Database analysis of ToxS offers little by way of suggestion for its precise biochemical function. Studies using a *ctx-lacZ* fusion in *E. coli* showed that ToxS is essential for ToxR activity at low levels of *toxR* gene expression, as when it is expressed from its own promoter in *E. coli*.[34] Consistent with the lack of homology between ToxR and OmpR-like response regulators in the phosphorylation domain, ToxS does not share similarity to sensor-kinase proteins. This lack of similarity between ToxR and ToxS and the other members of the two component family, as well as their membrane topology, suggests that the interaction between them is novel.

With one notable exception, strains of *V. cholerae* with mutations in the *toxS* gene do not express cholera toxin or other ToxR regulated virulence factors to levels as high as those seen in wild type strains. The exception is strain 569B, a historically well studied strain of *V. cholerae* that expresses high levels of cholera toxin and generally does so under conditions that are not permissive for its expression in most other strains.[7] This strain has a deletion at the *toxR* locus that removes *toxS* but not *toxR* (Mekalanos J, personal communication), and presumably a second mutation event has occurred that makes this strain independent of ToxS for cholera toxin production.

6. MOLECULAR ANALYSIS OF ToxR STRUCTURE AND FUNCTION

Alkaline phosphatase fusions to the periplasmic domain of ToxR have been used to investigate the structure and function of ToxR as well as to gain a better understanding of its periplasmic interaction with ToxS. Such proteins have the advantage over the wild type of expressing an assayable phenotype for activity occurring on both sides of the membrane. Transcriptional activation of *ctx* in *E. coli* and DNA binding offer an assessment of the function of the cytoplasmic domain of ToxR and alkaline phosphatase activity acts as a biochemical marker for the periplasmic domain of the protein.

Two ToxR-PhoA fusion proteins with differing amounts of ToxR periplasmic domain remaining in them demonstrated quite different responses when ToxS was expressed in the same cell.[33] When substantial amounts of the ToxR periplasmic domain were present in the fusion protein (ToxR-PhoA-L), ToxS was still required for activation of the *ctx-lacZ* promoter. In addition, while it was stable in *E. coli* in the presence of ToxS, this longer fusion protein was proteolytically cleaved in *E. coli* in the absence of ToxS. A different fusion protein, in which most of the periplasmic domain of ToxR was replaced with alkaline phosphatase, displayed quite different behavior. As indicated above, when all but six residues of the ToxR periplasmic domain were replaced by alkaline phosphatase, this fusion protein, called ToxR-PhoA S, was still able to bind and activate the *ctx* promoter. ToxR-PhoA-S was ToxS independent and not subject to the proteolysis observed with ToxR-PhoA-L. The most direct interpretation of these data is that the periplasmic domain of ToxR is required for the interaction between ToxR and ToxS and that the consequences of this interaction can be obtained by replacing the carboxy terminus of ToxR with alkaline phosphatase. In addition, the fact that ToxS could protect the longer ToxR-PhoA fusion protein from proteolysis in *E. coli* indicates that the interaction is intimate and may not be transient. These observations suggest a complex made up of ToxS and ToxR in which the latter has the capacity to bind to promoter DNA and activate transcription. It seems likely that the short fusion protein, which expresses alkaline phosphatase activity, is a dimer because active alkaline phosphatase is a dimer. In addition, because ToxR-PhoA-S expresses ToxR activity independent of ToxS, a model was proposed in which the active form of ToxR is a dimer and ToxS may play a role in dimer formation or stability.[33]

Other types of fusion proteins were constructed to analyze this model. These were designed to test whether ToxR requires dimerization at its carboxy terminus for activity and whether ToxR is capable of dimerizing. The yeast GCN4 leucine zipper (a dimerizing domain) and β-lactamase (a monomer) were used to replace the carboxy terminus of ToxR and both fusion proteins expressed ToxR activity (Ottemann K, PhD Thesis, Harvard University). There is an apparent requirement for membrane localization of ToxR, as a ToxR-GCN4 fusion protein lacking the transmembrane domain did not have ToxR activity (Ottemann K, and Mekalanos J, personal communication). These results suggest that transcriptional activation by ToxR does not require periplasmic dimerization for function.

Even though the results with ToxR-Bla and ToxR-GCN4 fusions described above indicate that periplasmic dimerization is not required for ToxR activity, the activity of another class of fusion protein suggests that nevertheless, ToxR has dimerization activity. In order to function, the DNA binding domain of lambda repressor (cI) requires dimerization. When fusion proteins were constructed in which this domain was fused at the amino terminus of ToxR, it bound its operator site, indicating that it had become dimerized.[35] Furthermore, this activity was enhanced in the presence of ToxS and, when the fusion was made in such a way as to replace all of the cytoplasmic domain of ToxR with cI, operator binding by the cI domain was absolutely dependent on ToxS. These results indicate that ToxR contains a dimerization domain and that ToxS either stabilizes spontaneously formed dimers or acts to recruit monomers into dimer formation.

That ToxR can dimerize but does not need to in order to function is the inescapable conclusion of all of these studies. Some of them were done using overexpressed fusion proteins, which may affect their interpretation. It may require that ToxR and ToxS be purified and their interaction with each other be analyzed in vitro before the most accurate conclusions regarding their interaction can be drawn.

The closest analogy to the ToxR/ToxS system in transcriptional regulatory systems described thus far is the potential interaction in *E. coli* between CadC, the transmembrane activator of the lysine decarboxylase operon and the lysine transporter, LysP, an integral membrane protein with as many 12 membrane spanning segments.[36] Maximum expression of lysine decarboxylase requires both low pH (5.8) and exogenously added lysine. Based on

the phenotypes of specific *lysP* and *cadC* mutant alleles, Olson and colleagues have proposed a model in which LysP and CadC may interact with one another in response to extracellular lysine concentrations.[37] In this model, unlike with the apparent enhancement of ToxR activity by interaction with ToxS, LysP is proposed to reduce CadC activity in the absence of lysine.

7. ANALYSIS OF ToxR DNA BINDING MUTANTS

Site directed mutagenesis showed that specific residues shared by ToxR and OmpR are critical for transcriptional activation and *ctx* promoter binding by ToxR.[30] Several residues were analyzed in this study, and arginine residues at positions 68, 77, and 96 (R68, R77, R96) were shown to be required either for activation of the *ctx* promoter or for *ctx* promoter DNA binding.[30] The magnitude of the effects of changes on *ctx-lacZ* activation varied from less than one percent of wild type activity for a mutant in which R96 was changed to lysine (ToxRR96K) or R77 to leucine (ToxRR77L), to four percent of wild type activity for ToxRR68K. Although ToxRR77L did not activate *ctx-lacZ* in *E. coli*, it could complement a *toxR* null mutant of *V. cholerae* for toxin production,[30] an observation that will be discussed below.

Two mutants that could not activate the *ctx-lacZ* promoter to any significant degree, ToxRR77L and ToxRR96L, retained the ability to shift the mobility of a DNA fragment containing the *ctx* promoter in gel electrophoresis.[30] DNA binding without transcriptional activation (a phenotype termed positive control), is associated with inability to interact productively with the alpha subunit of RNA polymerase and has been observed in mutants of other transcriptional activators.[38] Whether these residues are in fact important specifically for ToxR interaction with RNA polymerase could be clarified by, among other approaches, identifying suppressor mutations of these mutant ToxR protiens that map in the *rpoA* gene, which encodes the alpha subunit of polymerase. That ToxRR96K has neither DNA binding activity nor the ability to activate transcription seems to argue that if R96 is required for polymerase association, it may also play a role in DNA binding as well.

Another mutant phenotype worth noting is that of a protein in which the glutamic acid in position 51 was changed to lysine (ToxRE51K), which was dominant over the wild type in *V. cholerae*.[30] This phenotype suggests interaction between the mutant protein and the wild type protein and lends support to the model that ToxR may function as a multimer, although it could also be the result of competition between the mutant and the wild type protein for DNA sites to which ToxR binds. Irrespective of why this protein exhibits dominance over wild type, this was the only mutant, among 10 mutants tested, that did so.[30]

7.1. Promoter Elements Recognized by ToxR

The *ctx* and *toxT* promoters are the only ones that have been demonstrated to be direct substrates for ToxR binding.[29,31]

Based on the fact that *ompU* expression in *V. cholerae* requires ToxR,[28] but not ToxT (DiRita VJ, unpublished data) it might be that ToxR binds to and activates the *ompU* promoter also, but the gene has only recently been cloned[13] and its promoter structure has not been determined (Kaper JB, personal communication).

Sequence analysis of the *ctxAB* genes identified an element composed of directly repeated heptamers of the sequence TTTTGAT.[39] Band shift experiments with wild type and deletant promoter fragments demonstrated that these repeats are required for ToxR to bind to and activate the *ctx* promoter (Fig. 5.2).[29] Different strains of *V. cholerae* harbor different numbers of these repeats in front of the *ctx* genes, and it has been suggested that the number of repeats may correlate to the magnitude of cholera toxin expression.[39] If so, it may be that perhaps ToxR has a generally weak affinity for its binding site and the increase in the number of repeats allows cooperative binding of ToxR to the promoter.

The question of precisely how ToxR binds DNA is made more compelling with the identification of a ToxR binding site

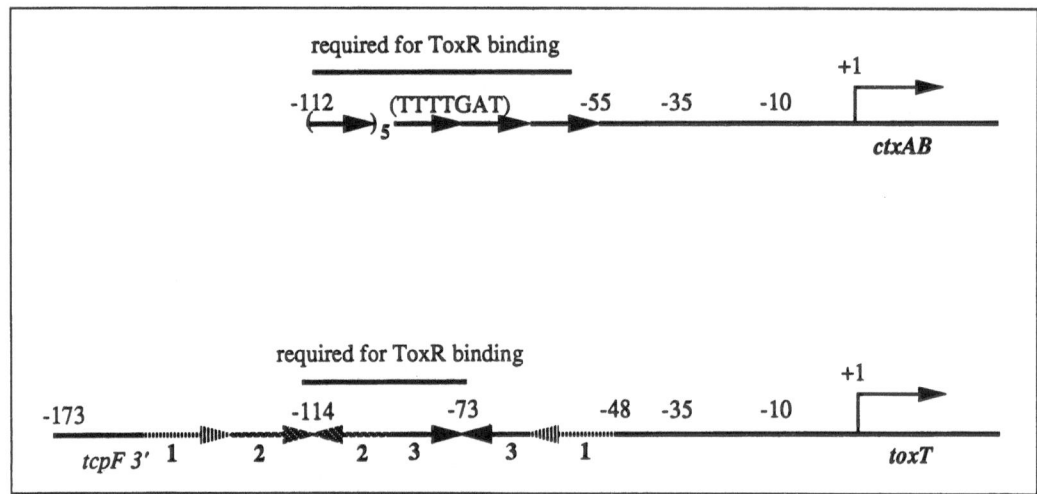

Fig. 5.2. The ctxAB and toxT promoters showing the structural elements identified by DNA sequence and transcriptional regulation analysis. The ctx promoter element, TTTTGAT, is repeated from three to eight times depending on the strain of V. cholerae, but deletion analysis was done with a strain harboring eight copies. The half sites for the inverted repeats in the toxT promoter are indicated by the type of arrow and by numbers. The regions required for ToxR binding are indicated by the bar above each promoter.

in the *toxT* promoter which has no evident similarity to that in the *ctx* promoter.[31] Deletion analysis of the *toxT* promoter identified a region of DNA between -73 and -114 (relative to the start site of transcription) that is required for ToxR dependent activation of *toxT* and for ToxR binding. DNA upstream of *toxT*, including the ToxR binding site, is rich in inverted repeat elements but does not contain any sequence element similar to the TTTTGAT heptamer repeat required for ToxR binding and activation of the *ctx* promoter (Fig. 5.2).[31] One of the repeat elements may act to terminate transcription of the upstream *tcpF* gene,[31] but repeat 3 (Fig. 5.2) is implicated in ToxR binding because deletions that remove half sites of repeats 1 and 2 are still bound by ToxR, while a deletion removing one of the half sites of repeat 3 is not. In addition, promoter mutations that abrogate ToxR interaction with the *toxT* promoter in *E. coli* map to repeat 3 (Higgins DE and DiRita VJ, unpublished data). These results indicate that ToxR binds specifically to two different sequence elements; in the *ctx* promoter, the ToxR binding site is defined by direct repeats and in the *toxT* promoter by inverted repeats (Fig. 5.2).

Although ToxR alone is able to activate the *ctx* promoter in *E. coli*, it cannot activate the *toxT* promoter in this background. On the contrary, when ToxR is expressed in an *E. coli* strain with a *toxT-lacZ* fusion on its chromosome, transcriptional activity from the fusion is decreased.[39] It is unlikely that a natural role for ToxR is to repress *toxT* expression, because *toxR* mutants express reduced, rather than elevated, levels of *toxT* mRNA.[31] Nevertheless, the ToxR dependent repression seen in *E. coli* requires the presence of the same sequences in the *toxT* promoter that are necessary for ToxR dependent activation of the promoter in *V. cholerae*. There are various reasons that might account for this observation. Perhaps the placement of the ToxR binding site in the *toxT* promoter may not enable ToxR to interact properly with the *E. coli* RNA polymerase, but does allow proper contact with the *V. cholerae* polymerase. Another possibility is that a *V. cholerae* factor, in addition to ToxR, is required to activate the bona fide *toxT* promoter and, absent this factor, ToxR binding to the *toxT* promoter represses transcription initiating from a cryptic promoter in *E. coli*.

The different effects of ToxR on the *ctx* and *toxT* promoters in *E. coli*, and whether this reflects differences in *V. cholerae*, may be further characterized by studying the mutant ToxR proteins described above. When tested for their ability to repress *toxT-lacZ* in *E. coli*, some of these mutant proteins behave predictably, based on their effect on *ctx* expression and promoter binding. That is, if the mutant has lost the ability to activate *ctx* it has also lost the ability to repress *toxT* in *E. coli*. Two proteins with potentially exceptional effects on the *toxT* promoter are ToxRR96L (arginine at residue 96 replaced with leucine) and ToxRR68K. ToxRR96L, which does not activate *ctx* but retains the wild type ability to bind the *ctx* promoter,[30] does not repress *toxT* expression in *E. coli*. ToxRR68K does not activate or bind to the *ctx* promoter,[30] yet represses *toxT* expression in *E. coli* with wild type efficiency (Higgins D and DiRita VJ, unpublished data). Further analysis of these two mutant proteins, to determine whether they bind the *toxT* promoter and whether they can complement a *toxR* mutant of *V. cholerae* for activation of the *toxT* promoter, will be important to understand the apparent difference in ToxR dependence by *ctx* and *toxT*.

Analysis of the effect of the ToxRR77L mutant protein on *toxT* expression may clarify why this mutant, which does not activate the *ctx-lacZ* fusion in *E. coli*, could still complement a *toxR* null mutant of *V. cholerae* for toxin production.[30] ToxR and ToxT both activate the cholera toxin promoter, and perhaps ToxRR77L cannot activate *ctx* directly, but still activates *toxT* expression in *V. cholerae*.

7.2. REGULATION OF
TOXT EXPRESSION BY TOXR

Of the virulence genes over which ToxR exerts regulatory control, it activates directly only the *ctx* operon.[40] The other genes require the ToxT transcriptional activator for their expression. Like ToxR, ToxT was identified in a screen of a *V. cholerae* plasmid library for factors that could activate a *ctx-lacZ* fusion in *E. coli*, based on genetic evidence from *V. cholerae* that such a factor should exist,[40] indicating that both activators recognize the *ctx* promoter (Fig. 5.1). Whether ToxT requires the same heptamer repeat element as ToxR for binding the *ctx* promoter has not been established.

When cloned *toxT* is expressed either from a constitutive[40] or inducible promoter (Bruss P and DiRita VJ, unpublished data), production of Ctx and TCP is restored in a *toxR* mutant strain of *V. cholerae*, leading to the model that ToxR activates *toxT* expression. This was confirmed with primer extension and *toxT-lacZ* fusion analyses, which, as described above, identified a ToxR specific promoter and binding site upstream of *toxT*.[31] OmpU production is not restored when *toxT* is overexpressed in a *toxR* mutant,[40] and is still expressed in a *toxT* mutant (DiRita VJ, unpublished data); these observations suggest that ToxR, but not ToxT, is required for *ompU* expression.

toxT is encoded within the *tcp* gene cluster, downstream of the *tcpF* gene.[17,41] In addition to ToxR dependent expression of the *toxT* promoter as described above, a larger transcript, perhaps originating at the ToxT dependent *tcpA* promoter and reading through a putative transcriptional terminator between *tcpF* and *toxT*, may also contribute to the total pool of *toxT* mRNA[31] (Taylor RK, personal communication). This raises the possibility that ToxT may activate transcription of a larger transcript that includes its own mRNA and thus be autoregulated. Notwithstanding the possibility that readthrough transcription all the way through the *tcp* cluster might generate *toxT* mRNA, *toxR* mutant *V. cholerae* that do not express *toxT* from the ToxR dependent promoter described above do not express ToxT regulated genes to any great extent.

8. TRANSCRIPTIONAL ACTIVATION BY TOXT

ToxT is a 32 kDa member of the AraC family of transcriptional regulatory proteins,[41] which is marked by sharing a conserved helix-loop-helix domain for DNA binding at the carboxy termini but lacking extensive similarity beyond that.[42] The mechanism for how ToxT activates gene expression is under investigation. Deletion analysis of the *acfA* promoter, which is activated to high levels by ToxT, identified a region in this promoter that is required for ToxT dependent activation (Peterson KM, personal communication). Purified ToxT, modified with six histidines at its amino terminus for purification purposes, binds to ToxT-dependent *acfA* promoter DNA, but not to deletants of this promoter that are not activated by ToxT (Bruss PM, Peterson KM, DiRita VJ; unpublished data). This suggests that ToxT controls gene expression by binding to promoter elements, although DNA sequence analysis of several ToxT dependent promoters does not identify any obvious sequence elements (direct or inverted repeats, for example) that ToxT might be interacting with.

One aspect of ToxT control may be the local topology of promoters with which it interacts. ToxR-mediated regulation of the *acfA* and *acfD* promoters in *V. cholerae* is abolished when the genes are expressed from a plasmid rather than from the chromosome, and is also abrogated in the presence of the DNA gyrase inhibitors nalidixic acid and novobiocin.[42] These observations suggest that activation of these promoters is sensitive to topological changes. DNA topology modulated by the HN-S protein plays a role in regulating gene expression controlled by CfaD, an AraC-like regulator of *E. coli*,[43] and perhaps this may be a feature of ToxT regulation as well.

9. SIGNAL RECOGNITION AND PROPAGATION IN THE ToxR REGULON

Implicit in the studies described above is that signals in the growth environment that influence toxin and other virulence factor expression are recognized by the ToxR system. The transmembrane structure of ToxR as well as its periplasmic interaction with ToxS suggest that signals might be recognized on the periplasmic side of the membrane, perhaps by both proteins, and be transmitted across the membrane to have a direct influence on ToxR regulated gene expression (Fig. 5.1). Below I describe some of the studies aimed at understanding how signals might be recognized by the ToxR system and the molecular aspects of the response to them.

9.1. OSMOLARITY

The initial observations that ToxR plays a role in coordinating gene expression in response to signals came from the study of ToxR-PhoA fusion proteins. Miller et al showed that a *toxR* mutant of *V. cholerae* could be complemented with a plasmid encoding a ToxR-PhoA fusion protein in which most of the periplasmic domain was replaced with alkaline phosphatase.[29] Cells expressing the fusion protein were able to express cholera toxin at both 66 mM and 500 mM NaCl, while wild type cells expressed toxin at 66 mM but not at 0 mM or 500 mM. The insensitivity to high osmolarity of cells expressing ToxR-PhoA suggests that the native ToxR may play a role in sensing the osmolarity of the surrounding medium and that the carboxy-terminal domain is responsible for this ability.

9.2. pH

Unlike the case with osmolarity, when ToxR-PhoA was expressed in a *toxR* mutant of *V. cholerae*, the response of the cells with respect to the pH effect on toxin production was parallel to that of wild type cells.[29] Thus, any pH sensing role of ToxR does not require the carboxy-terminus of the protein. Given that ToxR

and the *E. coli* pH sensitive activator CadC are most similar to one another in their cytoplasmic domains,[32] perhaps some of these shared residues are important for external pH sensing, although how this would be accomplished by an intracellular domain is not clear.

While the molecular mechanism of how ToxR may sense pH changes has yet to be worked out, the propagation of the pH signal is likely to proceed via activation of *toxT* transcription. This conclusion is based on the observation that *toxR* mutant *V. cholerae* can express TcpA and cholera toxin at both pH 6.5 and pH 8.5 if the *toxT* gene is expressed from a constitutive promoter, rather than from its own.[40] Consistent with this is that *toxT* transcription, assayed by primer extension, is higher at pH 6.5 than it is at pH 8.5 (Higgins D and DiRita VJ, unpublished data). A simple model is that under the permissive pH, ToxR activates *toxT* expression which leads to activation of *ctxAB* and the *tcp* genes. This model holds that, for pH at least, once ToxT is expressed no other pH specific signal is required for it to activate transcription.

9.3. TEMPERATURE AND HEAT SHOCK

That cholera toxin production is higher when cells are grown at lower temperature (30°C) than that of the human host (37°C) is unexpected. For example, *Yersinia* and *Shigella* spp., which are also intestinal pathogens, activate virulence factor expression in vitro at 37°C.[44,45] A model for how temperature might control ToxR regulated gene expression in *V. cholerae* developed from the observation that the *toxR* gene is divergently transcribed from the heat shock gene *htpG*, with fewer than 200 nucleotides between them.[46] Analysis of the activities of each promoter in *V. cholerae* grown at different temperatures showed that a proportional decrease in *toxR* expression and increase in *htpG* expression occurs over a range of temperature from 22°C to 37°C.[45] Thus, at 37°C, *toxR* expression was reduced five-fold over its expression at 22°C, while *htpG* expression was increased by a similar magnitude. In addition, simply expressing the RpoH heat shock sigma factor under *tac* promoter control in *E. coli* reduced expression of *toxT-lacZ* in *E. coli* irrespective of growth temperature.

To explain these data, Parsot and Mekalanos[46] suggested that RNA polymerases charged with either the vegetative sigma, σ^{70}, or the heat shock sigma, σ^{32}, compete for the overlapping *toxR* and *htpG* promoters. Under heat shock conditions, when σ^{32} levels increase, the model holds that a concomitant rise in RNA polymerase charged with σ^{32} occurs and *htpG* transcription increases at the expense of *toxR* transcription. This may occur, for instance, when the organisms first encounter the temperature of the human host after living in generally lower ambient temperatures outside of the host, or perhaps as a result of a host environment that induces the heat shock response. This model provides an explanation for how ToxR regulated gene expression may be kept low until the appropriate environment for the infecting organisms to produce colonization factors and toxin, without the requirement for specific transcription factors that control *toxR* expression.

9.4. TRANSMEMBRANE SIGNALING RESULTING FROM TOXR/TOXS INTERACTION

The phenotype of ToxR-PhoA-L, the long alkaline phosphatase fusion protein that retains wild type dependence on ToxS, was exploited to screen for mutants that would no longer interact productively with ToxS.[33] In the presence of ToxS the alkaline phosphatase activity of ToxR-PhoA-L was diminished and it was reasoned that if the ToxR/ToxS interaction were contributing to this decrease in alkaline phosphatase activity, mutants that express high alkaline phosphatase activity in the presence of ToxS might carry lesions in residues critical to the interaction.[33] Several mutants with this phenotype were identified and, in addition to their elevated alkaline phosphatase activity, these ToxS-blind

mutant ToxR-PhoA-L proteins were also subject to proteolytic cleavage in *E. coli* in the presence of ToxS, as the wild type protein is in the absence of ToxS.[33] Unexpectedly, these mutants carried lesions in the cytoplasmic domain of ToxR, at the distal end of the DNA binding domain. Introduction of the mutation into otherwise wild type *toxR* (i.e., not fused to *phoA*) abolished ToxR activity, indicating that these residues are critical for ToxR function.[33]

The domain identified by these mutations was termed the linker domain, and is conceptually similar to domains in two other membrane regulatory proteins, the BvgS membrane sensor protein controlling virulence gene expression in *Bordetella pertussis* and the methyl accepting chemotaxis protein Tsr of *E. coli*.[47,48] These domains are postulated to be involved in the transduction of signals that originate in the periplasm to the cytoplasmic domains of these transmembrane regulatory proteins.[47,48] In the case of ToxR, the linker domain may be critical for maintaining the structural features of ToxR that enable it to activate gene expression in response to ToxS. The linker domain mutants identified with the ToxR-PhoA-L screen described above may disrupt the active structure of ToxR and, in doing so, may destroy its interaction with ToxS.

10. CONCLUSIONS

The unique nature of the ToxR/ToxT regulatory cascade controlling virulence in *V. cholerae* raises many questions related to virulence and to general issues of gene regulation. One question worth further discussion here is how the *V. cholerae* regulatory cascade evolved. ToxR and ToxS are conserved in at least two other *Vibrio* species. In *V. parahaemolyticus*, they control transcription of a thermostable direct hemolysin[49] and the bioluminescent marine organism *V. fischeri* harbors both *toxR* and *toxS* homologues but the genes they control have yet to be identified.[50] A *toxT* homologue has not been identified in either of these species. The clustering of the *tcp-acf* genes (where the *toxT* gene resides) on the *V. cholerae* genome suggest that this large region might have been introduced into this species from another organism, perhaps by a phage or a plasmid. Upon introduction into *V. cholerae* these genes might over time, have been recruited into the ToxR system by evolution of the *toxT* promoter towards one that could be controlled by ToxR. The *ctx* genes, encoded on what appears to be a genetic element, might have come under ToxR control the same way. In this model, convergent evolution of the *ctx* and *toxT* promoters might have resulted in them both being controlled by ToxR, which might explain why they appear to be so different from one another. This model predicts the existence of other ToxR controlled genes in *V. cholerae*, which would have been in the postulated primordial strain prior to introduction of the *tcp-acf* cluster. A candidate for one of these would be *ompU*, which appears to be ToxT independent for its expression.

Many other questions regarding this system are of continuing research interest. For example, how does ToxR interact with the transcriptional machinery of the cell? Are there other such activators in bacteria (besides for CadC)? What role does the membrane spanning domain play for ToxR? What is the specific nature of the ToxR/ToxS interaction? How do culture signals impinge on this system? Does ToxT, like other AraC-like proteins, have a small molecule effector that controls its activity? Why is there regulatory redundance at the *ctx* promoter such that both ToxR and ToxT activate this operon? The novelty of this system, with its transmembrane transcriptional activator and apparent absence of a phosphorylation cascade of the type seen in so many other bacterial coordinate regulatory systems, is bound to continue to reveal interesting aspects of gene regulation in pathogenesis.

ACKNOWLEDGMENTS

I thank Ron Taylor, Ken Peterson, Jim Kaper, Karen Ottemann and John Mekalanos for sharing data prior to publication, and Sam Miller for his critical comments on a draft version of the manuscript.

REFERENCES

1. Goldberg MB, Boyko SA, Calderwood SB. Positive transcriptional regulation of an iron-regulated virulence gene in *Vibrio cholerae*. Proc Natl Acad Sci USA 1991; 88:1125-1129

2. Williams SG, Attridge SR, Manning PA. The transcriptional activator HlyU of *Vibrio cholerae*: nucleotide sequence and role in virulence gene expression. Mol Microbiol 1993; 9:751-760.

3. Gill DM. The arrangement of subunits in cholera toxin. Biochem 1976; 15:1242-1248.

4. Mekalanos JJ, Collier RJ, Romig WR. The enzymatic activity of cholera toxin. II. Relationships to proteolytic processing, disulfide bond reduction, and subunit composition. J Biol Chem 1979; 254:5855-5861

5. Gill DM, Meren R. ADP-ribosylation of membrane proteins catalyzed by cholera toxin: basis of activation of adenylate cyclase. Proc Natl Acad Sci USA 1978; 75:3050-3054.

6. Field M. Intestinal secretion and its stimulation by enterotoxins. In: Ouchterlony O, Holmgren J, eds. Cholera and related diarrheas. Karger: Basel, 1980: 46-52.

7. Richardson SH. Factors influencing in vivo skin permeability factor production by *Vibrio cholerae*. J Bacteriol 1969; 100:27-34.

8. Evans DJ, Richardson SH. In vitro production of choleragen and vascular permeability factor by *Vibrio cholerae*. J Bacteriol 1968; 96:126-130

9. Taylor RK, Miller VL, Furlong DB et al. Use of *phoA* gene fusions to identify a pilus colonization factor coordinately regulated with cholera toxin. Proc Natl Acad Sci USA 1987; 84:2833-2837.

10. Taylor RK. Genetic studies of enterotoxin and other potential virulence factors of *Vibrio cholerae*. In: Hopwood DA, Chater KF, eds. Genetics of bacterial di-

versity, London: Academic, 1989:309-329.

11. Peterson KM, Mekalanos JJ. Characterization of the ToxR regulon: identification of novel genes involved in intestinal colonization. Infect Immun 1988; 56:2822-2829.

12. Ogierman MA, Zabihi S, Mourtzios L et al. Genetic organization and sequence of the promoter-distal region of the *tcp* gene cluster of *Vibrio cholerae*. Gene 1993; 126:51-60.

13. Spanderio V, Giron JA, Silveira WD et al. Characterization of the outer membrane protein OmpU of *Vibrio cholerae*. American Society for Microbiology General Meeting, 1994; Abstract B-272

14. Herrington DA, Hall RH, Losonosky JJ et al. Toxin, toxin-coregulated pili, and the *toxR* regulon are essential for *Vibrio cholerae* pathogenesis in humans. J Exp Med 1988; 168:1487-1492.

15. Iredell JR, Manning PA. The toxin-coregulated pilus of *Vibrio cholerae* O1: a model for type 4 pilus biogenesis? Trends in Microbiol. 1994; 2:187-192.

16. Kaufman MR, Seyer JM, Taylor RK. Processing of TCP pilin by TcpJ typifies a common step intrinsic to a newly recognized pathway of extracellular protein secretion by gram-negative bacteria. Genes Develop 1991; 5:1834-1846

17. Kaufman MR, Shaw CE, Jones ID et al. Biogenesis and regulation of the *Vibrio cholerae* toxin-coregulated pilus: analogies to other virulence factor secretory systems and localization of the *toxT* virulence regulator to the TCP cluster. Gene 1993; 126:43-49.

18. Peek JA, Taylor RK. Characterization of a periplasmic thiol:disulfide interchange protein required for the functional maturation of secreted virulence factors of *Vibrio cholerae*. Proc Natl Acad Sci USA 1992; 89:6210-6214.

19. Yu J, Webb H, Hirst TR. A homologue of the *Escherichia coli* DsbA protein involved in disulfide bond formation is required for enterotoxin biogenesis in *Vibrio cholerae*. Mol Microbiol 1992; 6:1949-1958

20. Harkey CW, Everiss KD, Peterson KM. The *Vibrio cholerae* toxin-coregulated-pilus gene *tcpI* encodes a homologue of methyl-accepting chemotaxis proteins. Infect Immun 1994; 62: 2669-2678.

21. Freter R, O'Brien PCM. Role of chemotaxis in the association of motile bacterial with intestinal mucosa: fitness and virulence of nonchemotactic *Vibrio cholerae* mutants in infant mice. Infect Immun 1981; 34:222-233

22. Freter R, O'Brien PCM, Macsai MMS. Role of chemotaxis in the association of motile bacteria with intestinal mucosa: in vivo studies. Infect Immun 1981; 34:234-240

23. Mekalanos JJ. Duplication and amplification of toxin genes in Vibrio cholerae. Cell 1983; 35:253-263.

24. Fasano A, Baudry B, Pumplin DW et al. *Vibrio cholerae* produces a second enterotoxin which affects intestinal tight junctions. Proc Natl Acad Sci USA 1991; 88:5242-5246.

25. Trucksis M, Galen JE, Michalski J et al. Accessory cholera enterotoxin (Ace), the third toxin of a *Vibrio cholerae* virulence cassette. Proc Natl Acad Sci USA 1993; 90:5267-5271.

26. Pearson GDN, Woods A, Chiang SL et al. CTX genetic element encodes a site-specific recombination system and an intestinal colonization factor. Proc Natl Acad Sci USA 1993; 90:3750-3754

27. Miller VL, Mekalanos JJ. Synthesis of cholera toxin is positively regulated at the transcriptional level by *toxR*. Proc Natl Acad Sci USA 1984; 81:3471-3475.

28. Miller VL, Mekalanos JJ. A novel suicide vector and its use in construction of insertion mutations: osmoregulation of outer membrane proteins and virulence determinants in *Vibrio cholerae* requires *toxR*. J Bacteriol 1988; 170:2575-2583.

29. Miller VL, Taylor RK, Mekalanos JJ. Cholera toxin transcriptional activator ToxR is a transmembrane DNA binding protein. Cell 1987; 48:271-279.

30. Ottemann KM, DiRita VJ, Mekalanos JJ. ToxR proteins with substitutions in residues conserved with OmpR fail to acti-

vate transcription from the cholera toxin promoter. J. Bacteriol 1992; 174:6807-6814.

31. Higgins DE, DiRita VJ. Transcriptional control of *toxT*, a regulatory gene in the ToxR regulon of *Vibrio cholerae*. Mol Microbiol 1994; in press.

32. Watson N, Dunyak DS, Rosey EL et al. Identification of elements involved in transcriptional regulation of the *Escherichia coli cad* operon by external pH. J Bacteriol 1992; 174:530-540.

33. DiRita VJ, Mekalanos JJ. Periplasmic interaction between two membrane regulatory protein, ToxR and ToxS, results in signal transduction and transcriptional activation. Cell 1991; 64:29-37.

34. Miller VL, DiRita VJ, Mekalanos JJ. Identification of *toxS*, a regulatory gene whose product enhances ToxR-mediated activation of the cholera toxin promoter. J Bacteriol 1989; 171:1288-1293.

35. Dziejman M, Mekalanos JJ. Analysis of membrane protein interactions: ToxR can dimerize the amino terminus of phage lambda repressor. Mol Microbiol 1994; 13:485-494

36. Steffes C, Ellis J, Wu J et al. The *lysP* gene encodes the lysine-specific permease. J Bacteriol 1992; 174:3242-3249

37. Neely M, Dell CL, Olson ER. Roles of LysP and CadC in mediating the lysine requirement for acid induction of the *Escherichia coli cad* operon. J Bacteriol 1994; 176:3278-3285.

38. Russo FD, Silhavy TJ. Alpha: the cinderella subunit of RNA polymerase. J Biol Chem. 1992; 14515-14518.

39. Mekalanos JJ, Swartz DJ, Pearson GDN et al. Cholera toxin genes: nucleotide sequence, deletion analysis and vaccine development. Nature 1983; 306:551-557.

40. DiRita VJ, Parsot C, Jander G et al. Regulatory cascade controls virulence in *Vibrio cholerae*. Proc Natl Acad Sci USA 1991; 88:5403-5407.

41. Higgins DE, Nazareno E, DiRita VJ. The virulence gene activator ToxT from *Vibrio cholerae* is a member of the AraC family of transcriptional activators. J Bacteriol 1992; 174:6974-6980.

42. Parsot C, Mekalanos JJ. Structural analysis of the *acfA* and *acfD* genes of *Vibrio cholerae*: effects of DNA topology and transcriptional activators on expression. J Bacteriol 1992; 174:5211-5218.

43. Jordi BJAM, Dagberg B, de Haan LAM et al. The positive regulator CfaD overcomes the repression mediated by histone-like protein H-NS (H1) in the CFA/I fimbrial operon of *Escherichia coli*. EMBO J 1992; 11:2627-2632.

44. Maurelli AT, Blackmon B, Curtiss R III. Temperature-dependent expression of virulence genes in *Shigella* species. Infect Immun 1984; 43:195-201.

45. Cornelis G, Sluiters C, Lambert de Rouvroit C et al. Homology between VirF, the transcriptional activator of the *Yersinia virulence* regulon, and AraC, the *Escherichia coli* arabinose operon regulator. J Bacteriol 1989; 171:254-262.

46. Parsot C, Mekalanos JJ. Expression of ToxR, the transcriptional activator of the virulence factors in *Vibrio cholerae*, is modulated by the heat shock response. Proc Natl Acad Sci USA 1990; 87:9898-9902.

47. Miller JF, Johnson SA, Black WJ et al. Constitutive sensory transduction mutations in the *Bordetella pertussis bvgS* gene. J Bacteriol 1991; 970-980.

48. Ames P, Parkinson JS. Transmembrane signalling by bacterial chemoreceptors: E. coli transducers with locked signal output. Cell 1988; 55:817-826.

49. Lin Z, Kumagai K, Baba K et al. *Vibrio parahaemolyticus* has a homolog of the *Vibrio cholerae toxRS* operon that mediates environmentally induced regulation of the thermostable direct hemolysin gene. J Bacteriol. 1993; 175:3844-3855.

50. Reich KA, Schoolnik GK. The light organ symbiont *Vibrio fischeri* possesses a homolog of the *Vibrio cholerae* transmembrane transcriptional activator ToxR. J Bacteriol 1994; 176:3085-3088.

51. Parsot C, Mekalanos JJ. Expression of the *Vibrio cholerae* gene encoding aldehyde dehydrogenase is under control of ToxR, the cholera toxin transcriptional activator. J Bacteriol 1991; 2842-2851

ENVIRONMENTAL CONTROL OF VIRULENCE FUNCTIONS AND SIGNAL TRANSDUCTION IN *YERSINIA ENTEROCOLITICA*

Guy R. Cornelis, Maite Iriarte, and Marie-Paule Sory

1. INTRODUCTION

Among the many species of the *Yersinia* genus, only *Y. pestis*, *Y. pseudotuberculosis* and *Y. enterocolitica* adapted to multiply at the expenses of a host that is still alive. *Y. pestis* and *Y. pseudotuberculosis* are essentially rodent pathogens causing systemic diseases. *Y. enterocolitica* is a common human pathogen which causes gastrointestinal syndromes of various severities, ranging from mild self-limited diarrhea to mesenteric adenitis evoking an appendicitis. Although all three yersiniae invade their host via different routes, they share a common tropism for lymphoid tissue and a remarkable ability to resist the nonspecific immune response. Their main strategy seems to consist in avoiding lysis by complement and phagocytosis by polymorphonuclear leukocytes and macrophages and to form extracellular microcolonies in the infected tissue. Yersiniae succeed in infecting their host owing to the opportune production of a series of invasion and antihost proteins. The production of these proteins is tightly controlled by sophisticated regulatory networks: this rapidly ensures the survival of bacteria in hostile and changing environments. In *Yersinia*, genes encoding these proteins are either on the chromosome or distributed on a 70-kb plasmid called pYV, which is remarkably well conserved among the three species. In *Y. enterocolitica*, the chromosomal genes are mainly involved in the first steps of the infection while the pYV plasmid seems to be essentially devoted to resistance against the nonspecific immune response.

Signal Transduction and Bacterial Virulence, edited by Rino Rappuoli, Vincenzo Scarlato and Beatrice Aricò. © 1995 R.G. Landes Company.

We shall first describe the virulence functions and then focus on the regulation of their expression in response to environmental changes. For the sake of clarity, we will essentially deal with *Y. enterocolitica* and the differences with the other species will be mentioned throughout. For other reviews on yersiniae, see refs. 1-5.

2. CHROMOSOME-ENCODED VIRULENCE FUNCTIONS

2.1. THE ENTEROTOXIN YST

The chromosome of pathogenic *Y. enterocolitica* but not of *Y. pseudotuberculosis* and *Y. pestis* encodes a 30-amino acid, heat-stable, enterotoxin called Yst.[6] It resembles both the heat-stable enterotoxin ST_a (also called STI) of *E. coli* and guanylin, an endogenous activator of intestinal guanylate cyclase.[7] A study conducted with isogenic Yst+ and Yst- strains in the young rabbit concluded that, at least in this model, Yst is responsible for diarrhea. This suggests that Yst can also be responsible for diarrhea observed in young children infected with *Y. enterocolitica*.[8]

2.2. THE MYF FIBRILLAE

Y. enterocolitica synthesizes a fibrillar structure known as Myf and resembling the CS3 pili of human enterotoxigenic *E. coli*.[9] The assembly of Myf requires the classical components of the pili assembly systems, namely a periplasmic chaperone called MyfB and a channel-forming outer membrane protein called MyfC.[9] The counterpart of the *myf* operon in *Y. pestis* and *Y. pseudotuberculosis* encodes a structure known as pH6 antigen (pH6 Ag).[10] In *Y. enterocolitica*, we hypothesize that Myf could fulfill the role of a colonization factor for the human or porcine intestine, reinforcing the action of Yst, but this remains totally speculative.

2.3 THE INTERNALINS INV AND AIL

The chromosome of enteropathogenic *Y. enterocolitica* encodes two outer membrane proteins, Inv and Ail, mediating independent attachment followed by entry into cultured mammalian cells.[11,12] The signal transduction events occurring during these steps have been reviewed by Bliska et al.[13] In vivo, the invasin Inv appears to play a vital role in promoting entry in the intestinal tissue during the initial stage of the infection.[14]

3. pYV-ENCODED FUNCTIONS

3.1. YADA AND YLPA

YadA is a major outer membrane protein thought to form a fibrillar matrix on the surface of *Y. enterocolitica* and *Y. pseudotuberculosis*.[15,16] It is a polymer of 200 to 240 kDa formed by the association of approximately 50-kDa subunits addressed via the classical Sec export pathway.[16] YadA mediates a strong adherence to various substrates including mammalian cells. This tends to suggest that it may be a colonization factor, but some observations indicate that, at least in *Y. enterocolitica*, YadA is rather a major key in the defense against nonspecific immune responses. Indeed, it inhibits formation of the complement membrane attack complex and it prevents opsonization, which severely reduces phagocytosis and killing by polymorphonuclear leukocytes.[17,18] Because of a point mutation in the gene, *Y. pestis* does not produce YadA.[19]

YlpA is a 29-kDa outer membrane lipoprotein produced by *Y. enterocolitica* and *Y. pseudotuberculosis*.[20] The only element that pleads for a role in pathogenesis is of a genetic order: the expression of *ylpA* is regulated like that of the *yadA* and *yop* genes (see below).

3.2. THE YOP PROTEINS

The Yops are a set of 11 secreted proteins[21,22] synthesized in vivo in the course of the infection.[23,24]

The function of some of them is emerging. YopE is cytotoxic for cultured HeLa cells. It is responsible for depolymerization of actin microfilaments and it may be the major effector in the inhibition of phagocytosis of yersiniae by macrophages.[25] YopH is a protein tyrosine phosphatase[26] acting on multiple substrates in the cytoplasm of macrophages.[27] It contributes to the ability of *Yersinia* to resist phagocytosis by peritoneal macrophages[28] presumably by counteracting cell regulating kinases. The carboxy-terminal domain of YopH is similar to the catalytic domain of eukaryotic protein tyrosine phosphatases, which raises the appealing hypothesis that *yopH* could be of eukaryotic origin. The same seems to apply to YopO (YpkA),[29] a protein serine kinase, and to YopM. YopM is a 41-kDa protein sharing significant similarity with the domain of the α-chain of human platelet membrane glycoprotein Ib (GP1b) which binds thrombin and the von Willebrand factor.[30] YopM could prevent platelet-mediated host defense events such as the onset of the inflammatory response.

3.3. YOPS SECRETION

Yops are secreted by a new mechanism which does not involve the cleavage of a classical N-terminal signal sequence.[22] The addressing signal is nevertheless localized in the N-terminal domain of the protein.[31]

The Yop secretion machinery is encoded by the *virA*, *virB* and *virC* loci of the pYV plasmid (Fig. 6.1). The 8.5-kb *virC* locus constitutes a single large operon composed of 13 genes called *yscA* to *yscM* (for Yop secretion).[32] YscC belongs to the PulD family of outer membrane proteins which have representatives in different secretion pathways in Gram-negative bacteria (for a review see ref. 33). These proteins probably all form multimers, as recently shown for PIV, a member of the family involved in the extrusion of filamentous phages.[34] YscC would thus form a gated channel that conducts the Yops across the outer membrane. Homologues of the *yscJ*, *yscF* and *yscC* genes have been discovered in other animal pathogens such as *Shigella flexneri* and *Salmonella typhimurium* and also in plant pathogens such as *Xanthomonas campestris*, *Pseudomonas solanacearum* and *Pseudomonas syringae* (For a review see ref. 35).

The *virB* locus, called *lcrB* in *Y. pestis* and *Y. pseudotuberculosis*, consists of genes *yscN* to *yscU* arranged as a single operon.[36] YscN, the product of the first gene, is an ATP-binding protein that could act as an energizer of the system[37] while YscR and YscU are inner membrane proteins.[38,39] The *virB* locus, is the counterpart of the *spa* locus from *S. flexneri* and *S. typhimurium* (for reviews see refs. 3, 5). In addition, YscN has homologues in *X. campestris* and *P. solanacearum*,[37] and YscU has an homologue in *P. solanacearum*.[38] YscN, R, S, T, and U are also related to components of the flagellum assembly apparatus of *E. coli*, *S. typhimurium*, or *B. subtilis*. (For a review see ref. 5).

Apart from the *ysc* genes, Yop secretion also involves *lcrD*, a gene of the *virA* locus which encodes a 77-kDa inner membrane protein. Homologues of LcrD are involved in flagellar biogenesis in *Caulobacter crescentus*, in *Campylobacter jejuni*, in *S.typhimurium*, and in *B.subtilis* (see refs. in 40).

If one accepts that *virC* probably contains ten genes involved in Yop secretion, this pathway thus requires at least 19 genes excluding the *syc* genes encoding the individual chaperones (see below). Many of these 19 genes have counterparts in export systems from other bacterial pathogens and in flagellum assembly systems. The *Yersinia* Yop secretion pathway is thus a representative of a new secretion pathway essentially devoted to pathogenesis in animals and in plants. It probably derives from the system involved in the export of the flagellum components.

One of the peculiarities of the Yop secretion system is that it makes use of cytoplasmic chaperones that are specific for individual Yops.[41,42] The genes encoding chaperones SycE (specific Yop chaperone E) and SycH are localized next to the corresponding *yop* genes, namely *yopE* and *yopH*. If a *syc* gene is mutated, the corresponding Yop is no longer exported but secretion of the other Yops is not affected. We hypothesize that the role of these chaperones is to lead the nascent

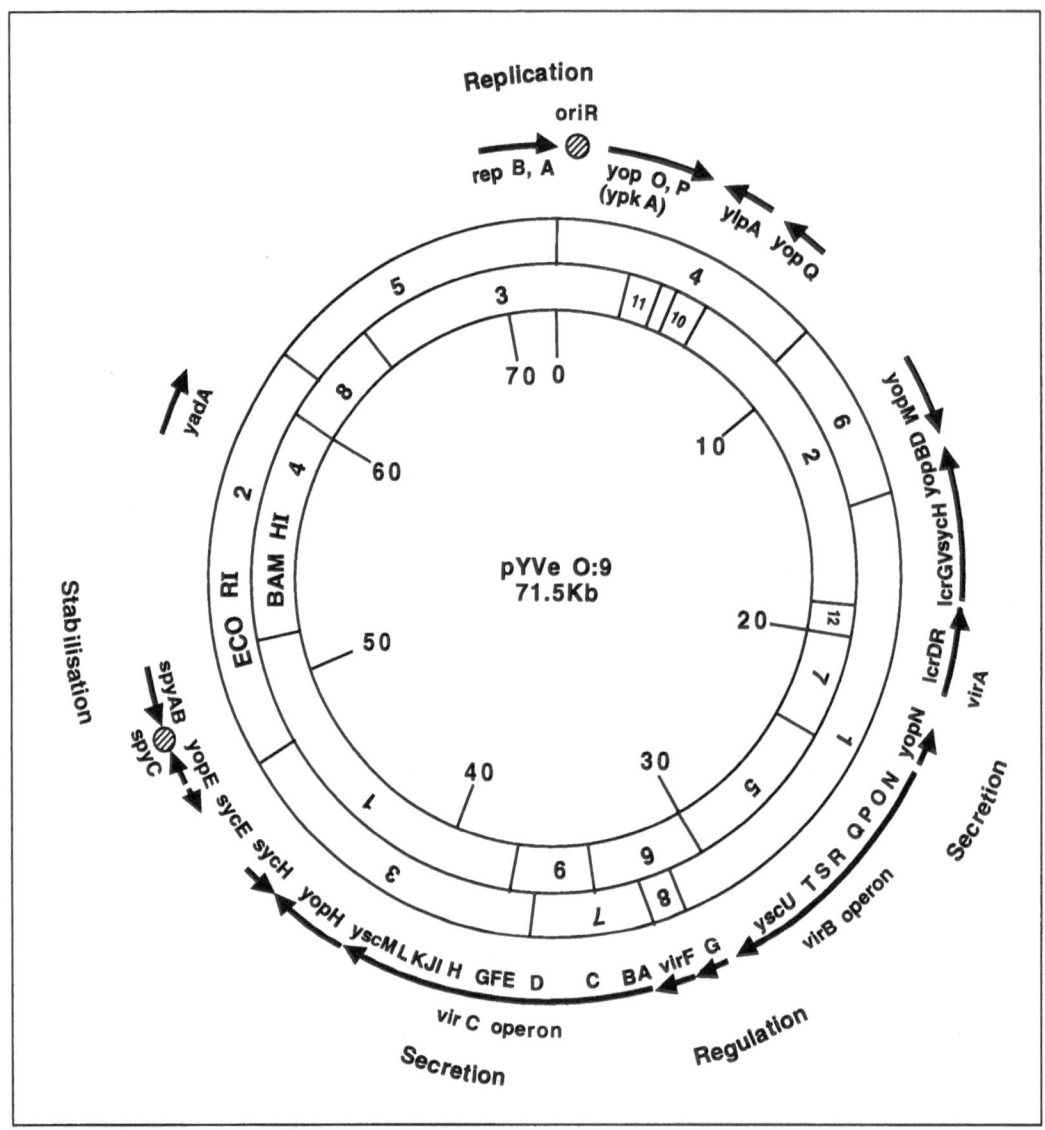

Fig. 6.1. The 70-kb pYV plasmid of a Y. enterocolitica strain of serotype O:9. Genes coding for antihost factors are scattered around the plasmid while temperature regulatory genes and secretion genes are assembled together in a ca. 20-kb region.

Yop proteins to the secretion apparatus. The Syc proteins would thus be the counterparts of SecB in the Sec-dependent pathway with the major difference that Syc proteins are specific for some individual Yops[41] while SecB is multivalent. SycE and SycH are very acidic proteins of 14.7 and 16 kDa respectively.[41,42] The chaperone serving YopB and YopD is LcrH, an acidic 18-kDa protein previously described as a gene regulator.[42]

3.4. TRANSLOCATION OF YOPS ACROSS THE MAMMALIAN CELL MEMBRANE

The presumed action of YopH as a phosphatase interfering with cellular regulatory kinases implies that it reaches the intracellular compartment. This also applies to cytotoxin YopE which disrupts the microfilament structure of cultured HeLa cells.[25,43] Cytotoxicity does not appear when purified YopE is added to cultured HeLa cells, but it appears when HeLa cells are infected by yersiniae adhering to the cell surface.[43] Recently, Rosqvist et al[44] showed by immunoprecipitation and fluorescence microscopy that YopE is translocated across the eukaryotic cell membrane to exert its toxic activity and that YopD plays a role in the translocation process.

Translocation of proteins by bacteria adhering to the cell surface is rather difficult to demonstrate formally. One has indeed to take into account the facts that bacteria adhering at the cell surface may contain Yops not yet secreted and that bacteria themselves can be internalized. Moreover, apart from YopH, the exact enzymatic activity of the translocated proteins is not known, which precludes the use of any sensitive enzymatic assay. We recently developed a reporter enzyme strategy to study translocation: we engineered a yersinia secreting a hybrid protein consisting of a Yop fused to a calmodulin-dependent adenylate cyclase and we reasoned that translocation into eukaryotic cells would result in an increase in cellular cAMP.[45] As expected, HeLa cells infected by the recombinant yersinia accumulate cAMP while cells infected by wild-type yersiniae do not. This increase in cAMP is strictly dependent on the presence of the adhesin YadA, the integrity of the secretion system and the production of an intact YopD (Fig. 6.2).

4. REGULATION OF CHROMOSOMAL VIRULENCE FUNCTIONS

Regulation of chromosomal virulence functions is summarized in Figure 6.3.

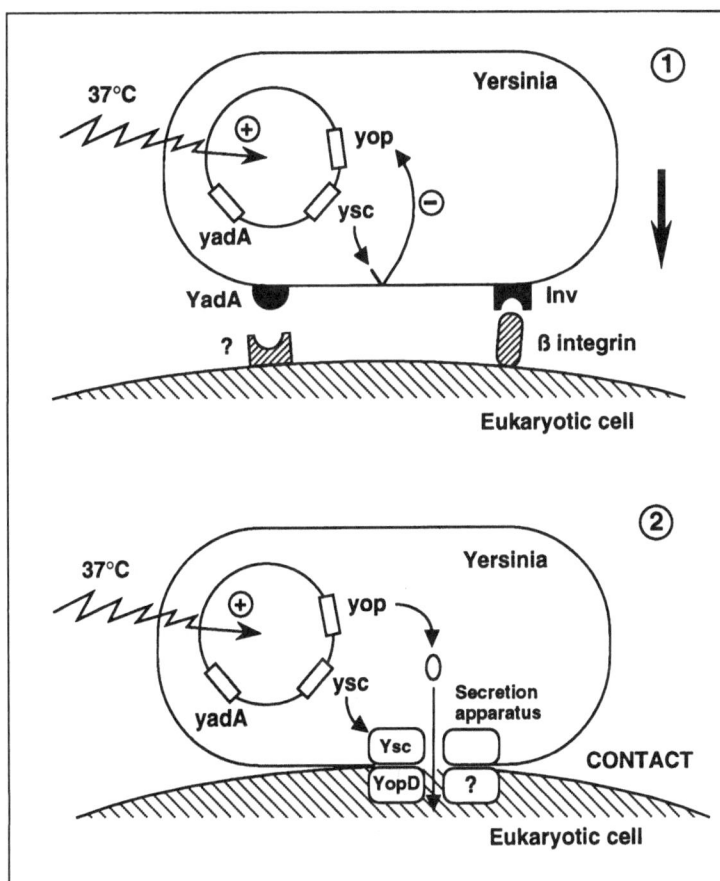

Fig. 6.2. Translocation of Yops from Y. enterocolitica into eukaryotic cells. Upon infection of mammalian cells at 37°C, yersiniae adhere to their surface by means of YadA and Inv. The resulting intimate contact endows secretion of Yops out of bacteria and their polarized entry in the cytoplasm of eukaryotic cells. If there is no contact with the eukaryotic cell, yop expression is inhibited through a feedback mechanism (see section 5.4 in the text). The complete process of translocation requires the secretion apparatus and an intact YopD protein.

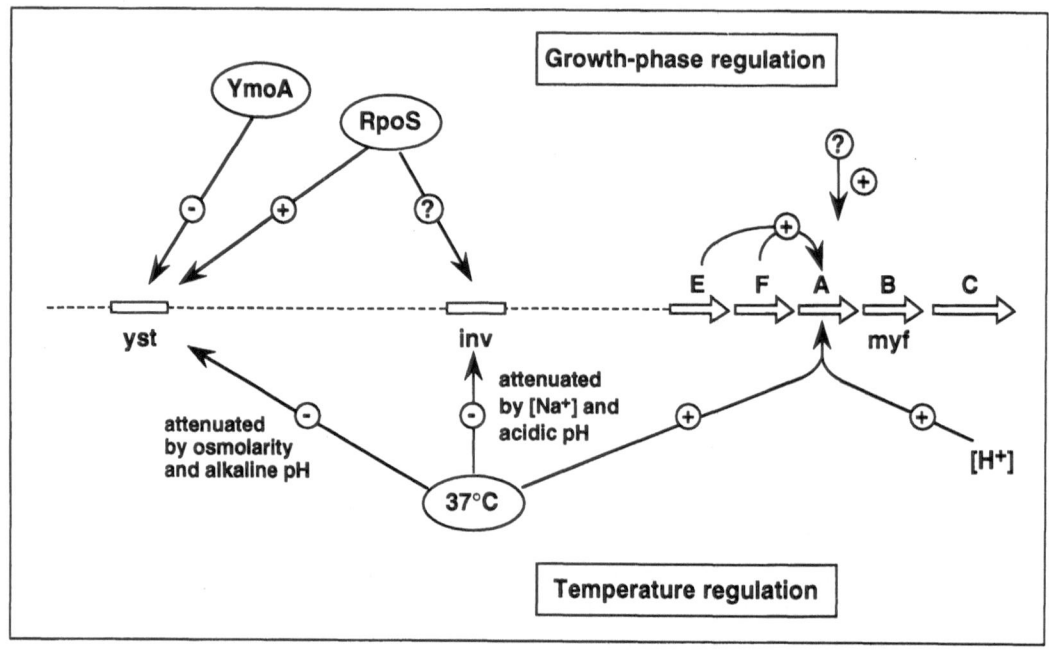

Fig. 6.3. Regulation of chromosomal virulence functions. Chromosomal genes are under the control of bacterial growth-phase and environmental parameters. Growth-phase regulation of yst involves RpoS and YmoA. The factor involved in growth phase regulation of myf has not been identified yet. Temperature regulation of inv and yst is modulated by Na⁺ and pH. myf transcription requires 37 °C, acidic pH, and the products of myfE and myfF.

4.1. GROWTH-PHASE REGULATION

The three chromosome-encoded virulence genes of *Y. enterocolitica*, namely *inv*, *yst*, and *myf* are growth-phase regulated, in the sense that none of them is expressed during the exponential growth. The fact that both Yst and Myf are produced at the same stage can be taken as an argument in favor of a coordinate role in vivo. There are however minor differences in the expression profiles. Expression of *inv* and *yst* begins when cells enter late-exponential phase to early stationary phase and continues throughout the stationary phase.[46,47] By contrast, *myfA* transcription occurs during transition between late-exponential phase and early stationary phase and then decreases gradually throughout the stationary phase.[47a]

Growth-phase dependent gene regulation is a highly complex phenomenon far from being completely understood. In *E. coli*, expression of many stationary-phase genes requires the alternative sigma factor RpoS (For a review see ref. 48). We have

recently isolated and mutagenized the homologue of *rpoS* in *Y. enterocolitica*.[47a] In such a mutant, *yst* expression is reduced though not completely abolished. The *Y. enterocolitica* homologue of RpoS is thus involved in the expression of *yst*[47a] but, as for many *E. coli rpoS*-regulated genes,[49,50] additional control mechanisms responding to growth-phase, must also be involved. In agreement with this, *yst* expression is modulated by osmolarity and by the histone-like protein YmoA (see below).[47] The mechanism of action of RpoS in *yst* regulation appears thus to be quite complex. So far, it is impossible to conclude whether RpoS regulates *yst* directly by binding to the promoter region, or indirectly through other regulatory proteins.

The role of *rpoS* in the expression of *inv* has not been established yet.[46] Finally, though, expression of *myf* occurs at the onset of stationary phase, is not significantly affected in a *rpoS* mutant.[47a] This type of situation is not unprecedented in *E. coli*.[50] *myf* and *inv* transcription also depends on temperature and pH (see below). These genes are thus under the dual control of bacterial growth phase and environmental factors.[46,47a,50a]

4.2. SILENCING DURING STORAGE

The level of *yst* expression depends on the bacterial strain analyzed: it gradually decreases in some strains during storage suggesting the existence of a mechanism switching the expression of *yst* to a silent state. The silencing of *yst* is not due to modifications in the gene itself but it rather reflects the status of bacterial host factors.[47] The production of Yst can be restored in a silent strain by an *ymoA* mutation,[51] but the typical growth phase dependent expression pattern is lost.[47] Taking into account the histone-like properties of Ymo (see below), chromatin structure could be involved in this gene reactivation phenomenon.

4.3 TEMPERATURE REGULATION

Temperature outside the mammalian host is generally less than 25°C while the host itself is thermally regulated at 37°C. It is not surprising that yersiniae, like several other animal pathogens, took advantage of this differential parameter to regulate the expression of their virulence panoply. All the yersinial pathogenicity functions are thermoregulated, though not always in the same direction (Fig. 6.3).

4.3.1. Temperature and ions control of *inv*

The host temperature of 37°C reduces transcription of *inv* as compared to ambient temperature.[52,53] Maximal expression at low temperature may prime *Y. enterocolitica* for cellular penetration once inside the host intestinal tract. After ingestion, *Y. enterocolitica* encounters a temperature of 37°C which could downregulate the production of Inv. However, temperature upshift is not the only stimulus which triggers *Y. enterocolitica* once they enter the host. Several environmental cues modulate thermoregulation and thus affect expression of *inv*. The most significant parameters in the intestinal tract are osmolarity, pH, oxygen and ions concentrations. In vitro, *inv* expression in *Y. enterocolitica* increases significantly at 37°C as pH of the growth medium goes below 7. *inv* expression at 37°C also increases with increasing concentration of Na^+.[46] Thus, in vivo, the acidity of the stomach, and the high Na^+ concentrations close to the enterocytes brush border could counteract the negative effect of temperature and allow continuous *inv* expression. In agreement with this hypothesis, Pepe et al[46] showed that the level of *inv* expression by *Y. enterocolitica* after 2 days in the mouse intestine is comparable to expression levels at 23°C in laboratory cultures.

4.3.2. Temperature and osmolarity control of *yst*

Production of Yst, like that of Inv, is also very low at 37°C in usual media.[6,47] This temperature downregulation of *yst* expression is quite puzzling because the enterotoxin needs to be secreted in vivo to induce diarrhea. This pattern of production recalls that of cholera toxin.[54] Moreover, as in the case of *inv*, thermoregulation of *yst* expression is modulated by additional factors such as osmolarity and pH: in vitro, *yst* transcription can be induced at 37°C by increasing Na^+ and K^+ ions concentrations as well as pH up to values normally present in the ileum lumen.[47] This observation reconciles the in vitro regulation studies with the known role of Yst in inducing diarrhea.

4.3.3. Temperature activation of *myf* transcription

Myf fibrillae are only produced at 37°C.[9] Myf production is thus downregulated when *Y. enterocolitica* are in the environment and it is stimulated when they enter their host. In agreement with this, we observed that mice infected with *Y. enterocolitica* grown at room temperature develop antibodies against Myf (Iriarte M, unpublished data). As for the other temperature regulated functions, temperature regulation of Myf production occurs at transcriptional level.[50a]

4.4. REGULATION OF MYF SYNTHESIS BY pH

Transcription of *myfA* requires a pH below 6.9.[9,50a] The same applies to pH6 Ag, the Myf counterpart in *Y. pestis*.[10,55] pH is not a parameter commonly regulating virulence functions however, it is also involved in the regulation of synthesis of Tcp pili by *Vibrio cholerae*.[56] This coincidence reinforces our hypothesis that Myf like Tcp, is a gut colonizing factor. After entering the host, Myf synthesis could be turned on by the acidic pH of the stomach. Nondividing bacteria might then keep their fibrillae in the intestine where alkaline pH prevails.

Tight regulation by acidity suggests that the regulating network controlling *myf* expression is more specific and probably more sophisticated that those controlling the expression of *yst* and *inv*. Transcription of *myfA* requires at least two genes, *myfF* and *myfE*, situated immediately upstream from *myfA*.[50a] Their products do not show similarity to any known regulatory protein. MyfF is a 18.5-kDa protein with no typical helix-turn-helix motif and an unique hydrophobic domain in the NH$_2$ terminal part. Hybrids between MyfF and PhoA constitute active phosphatases suggesting that MyfF is exported out of the cytoplasm. When produced in *E. coli* by the T7 promoter/polymerase system, MyfF is not released by an osmotic shock. We thus concluded that MyfF is associated with the inner membrane by means of its hydrophobic domain while the hydrophilic part protrudes in the periplasm. These features strikingly evoke ToxS, a protein involved in regulation of Tcp pilus production in *Vibrio cholerae*.[57]

Gene *myfE*, situated immediately upstream from *myf*, has not been characterized yet. In a *Y. enterocolitica myfE* mutant, transcription of *myfA* is completely abolished.[50a] The product of *myfE* presents significant identity to PsaE, a protein involved in regulation of pH6 Ag.[10,55]

Genes *myfF* and *myfE* are presumably part of a whole network because other chromosomal mutations, not yet analyzed, also affect expression of Myf (Iriarte M and Cornelis GR, unpublished data). This network is probably involved in regulation by acidity and, taking into account its membrane association, MyfF could be an element of the signal transducing system.

5. REGULATION OF pYV-ENCODED VIRULENCE FUNCTIONS

The pYV plasmid constitutes an integrated antihost device. It contains about 50 genes aligned without any gap and all devoted to adhesion and secretion of the Yop panoply (Fig. 6.1). Regulation of these pYV-encoded virulence genes is summarized in Figure 6.4.

5.1. VirF AND THE *virF* REGULON

Transcription of many genes among which all the *yop* genes, *sycE*, *ylpA*, *yadA* and the *virC* operon requires the product of *virF*, a gene located between *virB* and *virC*.[58] These genes and operons constitute what we call the *virF* regulon. By contrast, VirF seems to be dispensable for the transcription of the *virA* locus, of the *virB* operon and of some genes like *sycH*.[42,59] VirF is a 30-kDa transcriptional activator of the AraC family of regulators,[58] a very large family including regulators of degradative pathways in *E. coli* and *P. putida* as well as regulators involved in the control of virulence of *Shigella*, enterotoxigenic *E. coli* and the phytopathogen *P. solanacearum*.[60] VirF acts as a DNA-binding protein. DNaseI footprinting experiments on the VirF binding sites of four genes identified a protected region spanning about 40 bp immediately upstream from the RNA polymerase binding site. This VirF binding sequence is located in an AT-rich region and comprises two sites, each containing a 13-bp consensus sequence, either alone or invertedly repeated.[61]

5.2. THERMOREGULATION OF THE *YOP* STIMULON

The *yop*, *yadA* and *ylpA* genes as well as the *virA*, *virB* and *virC* loci are silent

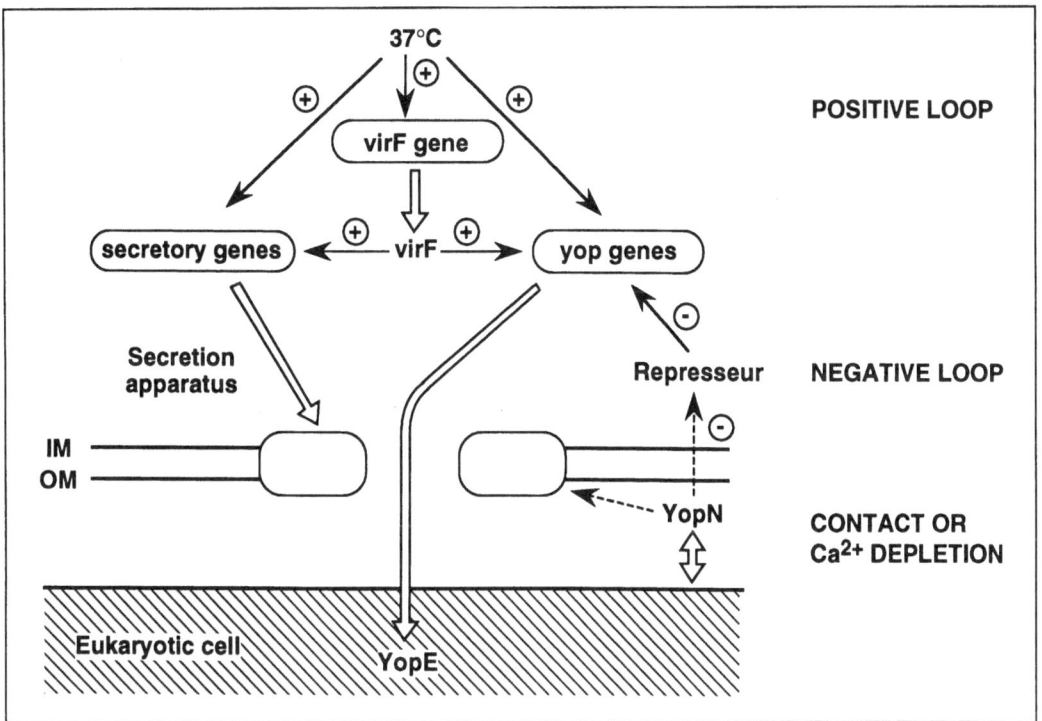

Fig. 6.4. *Temperature and Ca²⁺-or contact-regulatory networks for pYV-encoded functions. Positive loop: yop and secretory genes are switched on at 37 °C. Negative loop: without any contact with eukaryotic cells or in a medium containing Ca²⁺, a putative repressor reduces gene expression and abolishes Yops secretion. Upon contact with eukaryotic cells or in a medium deprived of Ca²⁺, the regulators of the negative loop are repressed by a mechanism involving YopN. The secretion apparatus is then assembled in the bacterial membranes at the zone of contact with the mammalian cell and Yops are translocated in its cytoplasm.*

at low temperature and strongly expressed at 37°C (see refs. in 58). They constitute what we call here the *yop* stimulon. Note that the *yop* stimulon is larger than the *virF* regulon.

In *Y. enterocolitica*, gene *virF* itself is strongly thermoregulated.[58] This thermo-induction still occurs in *E. coli* containing an isolated *virF* gene. Gene *virF* must thus be thermoregulated by a chromosomal gene rather than by the pYV plasmid. The fact that *virF* is thermoregulated can explain why the *virF* regulon is only expressed at 37°C. However, it does not demonstrate that the temperature regulation of the *virF* regulon only involves the temperature regulation of *virF*. Indeed, when *virF* is transcribed at low temperature from a *tac* promoter, the *yop* and *yadA* genes are only poorly transcribed.[59] By contrast, at 37°C, the response to IPTG

mimics the normal response to thermal induction.[59] In conclusion, the expression of the *yop* stimulon is first controlled by temperature but the expression of some genes is reinforced by the action of VirF, the synthesis of which is also temperature controlled.

In *Y. pestis* the situation is slightly different: transcription of *lcrF*, the homologue of *virF*, is insensitive to temperature. However, comparison of the amount of LcrF protein produced per unit of message at low and high temperatures indicates that the efficiency of translation of *lcrF* mRNA increases with temperature.[62] Based on sequence analysis, Hoe and Goguen[62] presented a model in which destabilization of an mRNA secondary structure sequestering the *lcrF* Shine-Dalgarno sequence regulates LcrF synthesis.

5.3. THE YMOA STORY

To identify the chromosomal regulator of the *yop* stimulon, we mutagenized a *Y. enterocolitica* strain which carries *lacZ* fused to *yopH*. We isolated two chromosomal mutants which strongly transcribe a *yopH-lacZ* gene fusion as well as *yopE* and *yadA* at 28°C. They nevertheless do not secrete the Yops at this temperature.[51] Transcription of the regulatory gene *virF* itself is increased at 28°C, which may account for the increased transcription of *yopH*, *yopE* and *yadA*. Although these genes are overexpressed at low temperature in the mutants, there is still an increase of transcription upon transfer to 37°C. Hence, the thermal response is not abolished in these two mutants but rather "modulated". The phenotype is thus not that of a classical repressor minus mutant. In both mutants, the transposon is inserted in the same gene called *ymoA* for "*Yersinia* modulator". The *ymoA* gene encodes a 8-kDa protein extremely rich in positively and negatively charged residues. Although there is no similarity between YmoA and HU, IHF or H-NS (H1), it is very likely that YmoA is a histone-like protein.[51] Chromatin structure is thus involved in the increasing susceptibility at 37°C, of the *yop* genes promoters to VirF activation. Temperature could somehow modify the structure of chromatin, making the promoters more accessible to VirF. Rohde et al[63] confirmed that temperature alters DNA supercoiling and DNA bending in *Y. enterocolitica* and they hypothesized that temperature dislodges a repressor, perhaps YmoA, bound on promoter regions of *virF* and of some other thermoregulated genes.

All the properties of the *ymoA* mutants are strikingly reminiscent of the *osmZ* mutants of *E. coli* and the *virR* mutants of *Shigella flexneri*, both lacking H-NS.[64-66] This observation is thus just another one pointing to the role of chromatin structure in transcription regulation. This phenomenon which seems now to be classical in thermoregulation of bacterial virulence functions is reviewed and discussed in detail by Dorman and Ni Bhriain.[67]

YmoA is not the *Yersinia* counterpart of H-NS from *E. coli* or *Shigella* but an homologue of YmoA was recently discovered in *E. coli*, as a regulator of hemolysin production.[68] The gene encoding this regulator, called *hha*, can complement the *ymoA* mutations of *Y. enterocolitica* provided it is expressed at the adequate level.[69] YmoA and Hha are thus the first representatives of a new class of histone-like proteins regulating the expression of topologically-sensitive promoters. Hha can thus be listed along with H-NS, IHF, FIS, HU and LRP. It is striking that the search for a thermoregulator, in *Yersinia*, in *Shigella* and in uropathogenic *E. coli*, converged on histone-like proteins.

5.4. CA²⁺-OR CONTACT REGULATION OF YOP SYNTHESIS AND SECRETION

Yops secretion in vitro requires not only a temperature of 37°C but also the absence of Ca^{2+} ions. By contrast, production of YadA is thermodependent but independent of the Ca^{2+} concentration. Hence, temperature and Ca^{2+} influence two different regulatory pathways. The first one, responding to temperature regulates all the pYV encoded virulence functions while the second one, responding to Ca^{2+}, only regulates the production of the Yops and YlpA. This second regulatory network is far from being perfectly understood yet. It has two aspects: (i) repression of *yop* genes transcription in non-permissive conditions for Yops secretion (presence of Ca^{2+}) and (ii) growth restriction in permissive conditions for Yops secretion (lack of Ca^{2+}).

5.4.1. The putative Ca²⁺ sensor and the hypothetical Ca²⁺ regulatory cascade

Yersiniae do not accumulate Ca^{2+} ions intracellularly. The negative influence of Ca^{2+} on expression or on secretion of Yops must thus involve one or several sensors or receptors expressed at the bacterial surface. Mutants affected in *yopN*, also called *lcrE*, are Ca^{2+} blind in the sense that they secrete Yops at 37°C even in the presence of Ca^{2+}.[70,71] In the model proposed by H. Wolf-Watz and his co-workers, YopN is deposited on the surface of the bacterium and acts either as a sensor for Ca^{2+} or in coordination with the Ca^{2+} sensor.[5,70] YopN has however never been shown to bind Ca^{2+}.

LcrG has also been postulated to be part of the Ca^{2+} sensing pathway.[72] A *lcrG* mutant is Ca^{2+} blind and fails to downregulate the secretion of at least YopM and LcrV. Like YopN, LcrG is a protein normally secreted by the Yop export apparatus. According to Skrzypek and Straley,[72] LcrG could act early in the negative regulatory cascade including YopN, because overexpression of LcrG in a *lcrE/yopN* mutant does not complement the mutation.

LcrQ could also act as a negative element in the Ca^{2+} regulatory pathway.[73] This protein, known as YscM in *Y. enterocolitica*, is encoded by the last gene of the *virC* operon.[32]

5.4.2. The contact signaling pathway for Yops secretion

The physiopathological significance of regulation by Ca^{2+} is far from being clear. Brubaker[74] suggested a correlation between Ca^{2+} dependency and the known, distinct levels of Ca^{2+} in the mammalian intracellular (µM range) and extracellular (about 2.5 mM) fluids. According to this suggestion, the Yops would essentially be produced in the intracellular environment. Although appealing, this hypothesis is contradicted by the accumulating evidence that yersiniae seem to spread and multiply essentially outside cells (see refs. in 3). There is thus a paradox: in vivo, yersiniae proliferate in conditions which are supposed to be nonpermissive for Yops production, but, yet, they do produce Yops.

Recently, YopE was shown to translocate from *Yersinia* into eukaryotic cells (see section 3.4).[44,45] This phenomenon which requires an intact secretion apparatus, is only realized by bacteria firmly attached to the surface of eukaryotic cells by Inv or YadA. Bacteria that are not in close contact with eukaryotic cells do not synthetize Yops.[44] Translocation of YopE

and of a hybrid YopE-cyclase into cells also occurs in a polarized manner: a mutant affected in translocation accumulates YopE or YopE-cyclase somewhere between the bacterium and the cytoplasm of the eukaryotic cell.[44,45] This translocation of YopE into cultured mammalian cells occurs in cell culture media containing about 1 mM Ca^{2+} and there is no release of Yops in the culture supernatant. Hence, secretion and translocation of YopE into mammalian cells occurs in a medium which is not permissive for Yop secretion. This suggests that in vivo, the phenomena are triggered by contact between bacteria and eukaryotic cells rather than by Ca^{2+} or another environmental signal. It remains that the Ca^{2+} regulation pathway is a key to study the factors involved in the contact signaling pathway. Indeed, YopN, predicted to be a Ca^{2+} sensor, in fact directs the entry of YopE inside HeLa cells. A *yopN* mutant is no longer polarized in the translocation of YopE and releases Yops in the medium surrounding HeLa cells.[44]

5.4.3. Feedback inhibition of Yops synthesis in absence of Yop secretion

Polar mutations in any of the *virA*, *B* and *C* loci encoding the secretion apparatus prevent the expression of the *yop* genes themselves.[58,75] This effect is maximal for mutations affecting the *virB* operon. Non-polar mutations inhibiting synthesis of the ATPase YscN or of the inner membrane protein YscU drastically reduce synthesis of the Yops.[37,38] This reveals the existence of a feedback inhibition mechanism of Yop synthesis when export is compromised. Its mode of action remains completely unknown. It could involve LcrQ.

5.4.4. Regulation of growth restriction

Although it is quite obvious that massive production of a given set of proteins must divert the metabolic potential from other biosynthetic pathways, it seems nevertheless that growth restriction associated with Yops secretion involves a specific inhibitory mechanism.

So far, growth restriction and secretion of Yops has always been coupled. Recently, we created a non-polar mutation in *virG*, a gene localized upstream from *virF* (Allaoui et al, submitted). A non-polar *virG* mutant grows at 37°C, independently of the Ca^{2+} concentration, but still produces and secretes all the Yops.

The phenotype of a *lcrV* mutant also argues for a role of LcrV in the growth restriction regulation.[76,77] LcrV is encoded by the second gene of the *lcrGVsycHyopBD* operon. A non-polar *lcrV* mutant does grow at 37°C, independently of the presence of Ca^{2+}, and to some extent, down-regulates transcription of some *yop* genes.

6. CONCLUSION

Like most of bacterial genes, virulence genes of yersiniae appear to be tightly regulated. The regulatory network involved is very complex and far from being completely understood yet. The major key is certainly temperature which influences any gene known to be involved in pathogenesis. Most of these genes are silent at 20°C and turned on at 37°C, with the exception of *inv* and *yst*, which are "on" at low temperature.

Like in uropathogenic *E. coli* and in *Shigella*, temperature regulation mainly involves topology of DNA and global regulators of the histone type. Such a global temperature regulation is expected to be modulated by other environmental parameters also known to influence DNA topology such as osmolarity and pH. This is particularly true for *inv* and *yst* which may thus be expressed in vivo at 37°C.

Specific regulatory systems are superimposed on this global temperature regulation. In particular, regulation of Myf fibrillae production by low pH probably involves a complete signal transduction cascade and a specific transcriptional regulator. Similarly, Yops secretion requires a classical transcription activator, VirF, which reinforces the effect of topological DNA modifications on some, though not all, genes

of the *yop* stimulon. Control by VirF is a rather classical control system and it is the best established regulatory pathway in yersiniae, though there seems to be variations between *Y. enterocolitica* and *Y. pestis*.

Yops secretion is repressed, in vitro, by the presence of Ca^{2+} ions. This second regulatory pathway, which is of the negative type, remains obscure. It remains obscure first with regard to its significance. Is Ca^{2+} the real parameter taken into account in vivo? The question is at least open since we know that yersiniae inject Yops into eukaryotic cells cultured in Ca^{2+}-containing media. Contact between the bacterial outer membrane and the mammalian cell membrane could rather be the actual signal recognized in vivo. The so-called Ca^{2+} negative regulatory loop also remains mysterious with regard to its mechanism. No classical repressor has indeed been identified so far but genetic observations argue in favor of some candidates. The negative loop must also involve a Ca^{2+}- or contact-sensor and a signal transduction cascade. YopN must somehow participate in these events but at what stage? The negative loop might well become clear soon with the current characterization of the secretion system. There is indeed a strong feedback inhibitory mechanism acting on Yop synthesis when secretion is compromised by a defect in the apparatus. This feedback inhibition could well be related to the Ca^{2+}- or contact-regulatory loop. The hypothesis of a secretable inhibitor, like that discovered in the control of flagellum synthesis[78] is very appealing but could not be demonstrated so far.

ACKNOWLEDGMENTS

Our research is supported by grant Nr 3.4627.93 from the Belgian Fund for Medical Scientific Research (FRSM) and by an "Action de Recherche concertée" granted by the GD Higher Education and Scientific Research, French Community of Belgium.

REFERENCES

1. Brubaker RR. Factors promoting acute and chronic diseases by Yersiniae. Clin Microbiol Rev 1991; 4:309-324.

2. Cornelis GR. Yersiniae, finely tuned pathogens. In: Hormaeche C, Penn CW, Smyth CJ, eds. Molecular biology of bacterial infection: current status and future perspectives. Cambridge University Press, 1992; 49:231-265.

3. Cornelis GR. *Yersinia* pathogenicity factors. Curr Topics Microbiol Immunol 1994; 192:243-263.

4. Straley SC, Plano GV, Skrzypek E et al. Regulation by Ca²⁺ in the *Yersinia* low - Ca²⁺ response. Mol Microbiol 1993; 8:1005-1010.

5. Forsberg A, Rosqvist R, Wolf-Watz H. Regulation and polarized transfer of the *Yersinia* outer proteins (Yops) involved in antiphagocytosis. Trends Microbiol 1994; 2:14-19.

6. Delor I, Kaeckenbeek A, Wauters G et al. Nucleotide sequence of *yst*, the *Yersinia enterocolitica* gene encoding the heat-stable enterotoxin, and prevalence of the gene among the pathogenic and non-pathogenic yersiniae. Infect Immun 1990; 58:2983-2988.

7. Currie MG, Fok KF, Kato J et al. Guanylin: an endogenous activator of intestinal guanylate cyclase. Proc Natl Acad Sci USA 1992; 89:947-951.

8. Delor I, Cornelis GR. Role of *Yersinia enterocolitica* Yst toxin in experimental infection of young rabbits. Infect Immun 1992; 60:4269-4277.

9. Iriarte M, Vanooteghem JC, Delor I et al. The Myf fibrillae of *Yersinia enterocolitica*. Mol Microbiol 1993; 9:507-520.

10. Lindler LE, Tall BD. *Yersinia pestis* pH 6 antigen forms fimbriae and is induced by intracellular association with macrophages. Mol Microbiol 1993; 8:311-324.

11. Isberg RR, Falkow S. A single genetic locus encoded by *Yersinia pseudotuberculosis* permits invasion of cultured animal cells by *Escherichia coli* K-12. Nature 1985; 317:262-264.

12. Miller VL, Falkow S. Evidence for two genetic loci in *Yersinia enterocolitica* that can promote invasion of epithelial cells. Infect Immun 1988; 56:1242-1248.

13. Bliska JB, Galan JE, Falkow S. Signal transduction in the mammalian cell during bacterial attachment and entry. Cell 1993; 73:903-920.

14. Pepe JC, Miller VL. *Yersinia enterocolitica* invasin: a primary role in the initiation of infection. Proc Natl Acad Sci USA 1993; 90:6473-7477.

15. Kapperud G, Namork E, Skurnik M et al. Plasmid-mediated surface fibrillae of *Yersinia pseudotuberculosis* and *Yersinia enterocolitica*: relationship to the outer membrane protein YOP1 and possible importance for pathogenesis. Infect Immun 1987; 55:2247-2254.

16. Skurnik M,Wolf-Watz H. Analysis of the *yopA* gene encoding the Yop1 virulence determinants of *Yersinia* spp. Mol Microbiol 1989; 3:517-529.

17. China B, Sory MP, N'Guyen BT et al. Role of the YadA protein in prevention of opsonization of *Yersinia enterocolitia* by C3b molecules. Infect Immun 1993; 61:3129-3136.

18. China B, N'Guyen BT, De Bruyere M et al. Role of YadA in resistance of *Yersinia enterocolitica* to phagocytosis by human polymorphonuclear leukocytes. Infect Immun 1994; 62:1275-1281.

19. Rosqvist R, Skurnik M, Wolf-Watz H. Increased virulence of *Yersinia pseudotuberculosis* by two independent mutations. Nature 1988; 334:522-525.

20. China B, Michiels T, Cornelis GR. The pYV plasmid of *Yersinia* encodes a lipoprotein YlpA, related to TraT. Mol Microbiol 1990; 9:1585-1593.

21. Heesemann J, Gross U, Schmidt N et al. Immunochemical analysis of plasmid-encoded proteins released by enteropathogenic *Yersinia* sp. grown in calcium-deficient media. Infect Immun 1986; 54:561-567.

22. Michiels T, Wattiau P, Brasseur R et al. Secretion of Yop proteins by yersiniae.

Infect Immun 1990; 58:2840-2849.

23. Martinez RJ. Plasmid-mediated and temperature-regulated surface properties of *Yersinia enterocolitica*. Infect Immun 1983; 41:921-930.

24. Sory MP, Cornelis GR. *Yersinia enterocolitica* O:9 as a potential live oral carrier for protective antigens. Microb Pathog 1988; 4:431-442.

25. Rosqvist R, Forsberg A, Wolf-Watz H. Intracellular targeting of the *Yersinia* YopE cytotoxin in mammalian cells induces actin microfilament disruption. Infect Immun 1991; 59:4562-4569.

26. Guan K, Dixon JE. Protein tyrosine phosphatase activity of an essential virulence determinant in *Yersinia*. Science 1990; 249:553-556.

27. Bliska JB, Guan K, Dixon JE et al. A mechanism of bacterial pathogenesis: tyrosine phosphate hydrolysis of host proteins by an essential *Yersinia* virulence determinant. Proc Natl Acad Sci USA 1991; 88:1187-1191.

28. Rosqvist R, Bölin I, Wolf-Watz H. Inhibition of phagocytosis in *Yersinia pseudotuberculosis*: a virulence plasmid-encoded ability involving the Yop2b protein. Infect Immun 1988; 56:2139-2143.

29. Galyov EE, Hakansson S, Forsberg A et al. A secreted protein kinase of *Yersinia pseudotuberculosis* is an indispensable virulence determinant. Nature 1993; 361: 730-732.

30. Reisner BS, Straley SC. *Yersinia pestis* YopM: Thrombin binding and overexpression. Infect Immun 1992; 60:5242-5252.

31. Michiels T, Cornelis GR. Secretion of hybrid proteins by the *Yersinia* Yop export system. J Bacteriol 1991; 173:1677-1685.

32. Michiels T, Vanooteghem JC, Lambert de Rouvroit C et al. Analysis of *virC*, an operon involved in the secretion of Yop proteins by *Yersinia enterocolitica*. J Bacteriol 1991; 173:4994-5009.

33. Genin S, Boucher CA. A superfamily of proteins involved in different secretion pathways in Gram-negative bacteria: modular structure and specificity of the N-terminal domain. Mol Gen Genet 1994; 243:112-118.

34. Kazmierczak BI, Mielke DL, Russel M et al. PIV, a filamentous phage protein that mediates phage export across the bacterial cell envelope, forms a multimer. J Mol Biol 1994; 238:187-198.

35. Van Gijsegem F, Genin S, Boucher CA. Conservation of secretion pathways for pathogenicity determinants of plant and animal pathogenic bacteria. Trends Microbiol 1993; 1:175-180.

36. Bergman T, Erickson K, Galyov E et al. The *lcrB* (*ysc* N/U) gene cluster of *Yersinia pseudotuberculosis* is involved in Yop secretion and shows high homology to the *spa* gene clusters of *Shigella flexneri* and *Salmonella typhimurium*. J Bacteriol 1994; 176:2619-2626.

37. Woestyn S, Allaoui A, Wattiau P et al. YscN, the putative energizer of the *Yersinia* yop secretion machinery. J Bacteriol 1994; 176:1561-1569.

38. Allaoui A, Woestyn S, Sluiters C et al. YscU, a *Yersinia enterocolitica* inner membrane protein involved in Yop secretion. J Bacteriol 1994; 176:4534-4542.

39. Fields KA, Plano GV, Straley SC. A low-Ca^{2+} response (LCR) secretion (ysc) locus lies within the *lcrB* region of the LCR plasmid in *Yersinia pestis*. J Bacteriol 1994; 176:569-579.

40. Plano GV, Straley SC. Multiple effects of *lcrD* mutations in *Yersinia pestis*. J Bacteriol 1993; 175:3536-3545.

41. Wattiau P, Cornelis GR. SycE, a chaperone-like protein of *Yersinia enterocolitica* involved in the secretion of YopE. Mol Microbiol 1993; 8:123-131.

42. Wattiau P, Bernier B, Deslée P et al. Individual chaperones required for Yop secretion by *Yersinia*. Proc Natl Acad Sci USA 1994; 91:10495-10497.

43. Rosqvist R, Forsberg A, Rimpilainen M et al. The cytotoxic protein YopE of *Yersinia* obstructs the primary host defence. Mol Microbiol 1990; 4:657-667.

44. Rosqvist R, Magnusson KE, Wolf-Watz H. Target cell contact triggers expression and polarized transfer of *Yersinia* YopE cytotoxin into mammalian cells. EMBO

J 1994; 13:964-972.

45. Sory MP, Cornelis GR. Translocation of a hybrid YopE-adenylate cyclase from *Yersinia enterocolitica* into HeLa cells. Mol Microbiol 1994; 14:583-594.

46. Pepe JC, Badger JL, Miller VL. Growth phase and low pH affect the thermal regulation of the *Yersinia enterocolitica inv* gene. Mol Microbiol 1994; 11:123-135.

47. Mikulskis AV, Delor I, Ha Thi V et al. Cornelis GR. Regulation of the *Yersinia enterocolitica* enterotoxin Yst gene. Influence of growth phase, temperature, osmolarity, pH and bacteria host factors. Mol Microbiol 1994; 14:905-915.

47a. Iriarte M, Stainier I, Cornelis GR. The rpoS gene from *Yersinia enterocolitica* and its influence on expression of virulence factors. Infect Immun 1995; in press.

48. Hengge-Aronis R. Survival of hunger and stress: the role of *rpoS* in early stationary phase gene regulation in *E. coli*. Cell 1993;72:165-168.

49. Lange R, Barth M, and Hengge-Aronis R. Complex transcriptional control of the σ^S-dependent stationary-phase-induced and osmotically regulated *osmY* (*csi-5*) gene suggests novel roles for Lrp, cyclic AMP (cAMP) receptor protein-cAMP complex, and integration host factor in the stationary-phase response of *Escherichia coli*. J Bacteriol 1993; 175:7910-7917.

50. Weichart D, Lange R, Henneberg N et al. Identification and characterization of stationary phase-inducible genes in *Escherichia coli*. Mol Microbiol 1993; 10:407-420.

50a. Iriarte M, Cornelis GR. MyfF, an element of the network regulating the synthesis of fibrillae in *Yersinia enterocolitica*. J Bacteriol 1995; 177(3):738-744.

51. Cornelis GR, Sluiters C, Delor I et al. *ymoA*, a *Yersinia enterocolitica* chromosomal gene modulating the expression of virulence functions. Mol Microbiol 1991; 5:1023-1034.

52. Isberg RR, Swain A, Falkow S. Analysis of expression and thermoregulation of the *Yersinia pseudotuberculosis inv* gene with

hybrid proteins. Infect Immun 1988; 56:2133-2138.

53. Pierson DE, Falkow S. Nonpathogenic isolates of *Yersinia enterocolitica* do not contain functional *inv*-homologous sequences. Infect Immun 1990; 58:1059-1064.

54. Parsot C, Mekalanos JJ. Expression of ToxR, the transcriptional activator of the virulence factors in *Vibrio cholerae*, is modulated by the heat shock response. Proc Natl Acad Sci USA 1990; 87:9898-9902.

55. Lindler LE, Klempner MS, Straley S. *Yersinia pestis* pH 6 antigen: genetic, biochemical, and virulence characterization of a protein involved in the pathogenesis of bubonic plague. Infect Immun 1990; 58:2569-2577.

56. Taylor RK, Miller VL, Furlong DB, Mekalanos JJ. Use of *phoA* gene fusions to identify a pilus colonization factor coordinately regulated with cholera toxin. Proc Natl Acad Sci USA 1987; 84:2833-2837.

57. DiRita VJ, Mekalanos JJ. Periplasmic interaction between two membrane regulatory proteins, ToxR and ToxS, results in signal transduction and transcriptional activation. Cell 1991; 64:29-37.

58. Cornelis GR, Sluiters C, Lambert de Rouvroit C et al. Homology between VirF, the transcriptional activator of the *Yersinia* virulence regulon, and AraC, the *Escherichia coli* arabinose operon regulator. J Bacteriol 1989; 171:254-262.

59. Lambert de Rouvroit C, Sluiters C, Cornelis GR. Role of the transcriptional activator VirF and temperature in the expression of the pYV plasmid genes of *Yersinia enterocolitica*. Mol Microbiol 1992; 6:395-409.

60. Genin S, Gough CL, Zischek C et al. Evidence that the *hrpB* gene encodes a positive regulator of pathogenicity genes from *Pseudomonas solanacearum*. Mol Microbiol 1992; 6:3065-3076.

61. Wattiau P, Cornelis GR. Identification of DNA sequences recognized by VirF, the transcriptional activator of the *Yersinia yop* regulon. J Bacteriol 1994;

176:3878-3884

62. Hoe NP, Goguen JD. Temperature sensing in *Yersinia pestis* : translation of the LcrF activator protein is thermally regulated. J Bacteriol 1993; 175:7901-7909.

63. Rohde JR, Fox JM, Minnich SA. Thermoregulation in *Yersinia enterocolitica* is coincident with changes in DNA supercoiling. Mol Microbiol 1994; 12:187-199.

64. Higgins CF, Dorman CJ, Stirling DA et al. A physiological role for DNA supercoiling in the osmotic regulation of gene expression in *S. typhimurium* and *E. coli*. Cell 1988; 52:569-584.

65. Dorman CJ, Bhriain NN, Higgins CF. DNA supercoiling and environmental regulation of gene expression in *Shigella flexneri*. Nature 1990; 344:789-792.

66. Göransson M, Sonden B, Nilsson P et al. Transcriptional silencing and thermoregulation of gene expression in *Escherichia coli*. Nature 1990; 344:682-685.

67. Dorman CJ, Bhriain NN. DNA topology and bacterial virulence gene regulation. Trends Microbiol 1993; 1:92-99.

68. Nieto JM, Carnona M, Bolland S et al. The *hha* gene modulates haemolysin expression in *Escherichia coli*. Mol Microbiol 1991; 5:1285-1293.

69. Mikulskis AV, Cornelis GR. A new class of proteins regulating gene expression in enterobacteria. Mol Microbiol 1994; 11:77-86.

70. Forsberg A, Vitanen AM, Skurnik M et al. The surface-located YopN protein is involved in calcium signal transduction in *Yersinia pseudotuberculosis*. Mol Microbiol 1991; 5:977-986.

71. Yother J, Goguen JD. Isolation and characterization of Ca^{2+} blind mutants in *Yersinia pestis*. J Bacteriol 1985; 164:704-711.

72. Skrzypek E, Straley SC. LcrG, a secreted protein involved in negative regulation of the low-calcium response in *Yersinia pestis*. J Bacteriol 1993; 175:3520-3528.

73. Rimpilainen M, Forsberg A, Wolf-Watz H. A novel protein, LcrQ, involved in the low-calcium response of *Yersinia pseudotuberculosis* shows extensive homology to YopH. J Bacteriol 1992; 174:3355-3363.

74. Brubaker RR. The Vwa^+ virulence of *Yersinia*: the molecular basis of the attendant nutritional requirement for Ca^{++}. Rev Infect Dis 1983; 5:5748-5758.

75. Cornelis GR, Vanooteghem JC, Sluiters C. Transcription of the *yop* regulon from *Y. enterocolitica* requires trans acting pYV and chromosomal genes. Microb Pathog 1987; 2:367-379.

76. Price SB, Cowan C, Perry RD et al. The *Yersinia pestis* V antigen is a regulatory protein necessary for Ca^{2+}-dependent growth and maximal expression of low-Ca^{2+} response virulence genes. J Bacteriol 1991; 173:2649-2657.

77. Bergman T, Hakansson S, Forsberg A et al. Analysis of the V antigen *lcr*GvH-*yop*BD operon of *Yersinia pseudotuberculosis*: evidence for a regulatory role of LcrH and LcrV. J Bacteriol 1991; 173:1607-1616.

78. Hughes KT, Gillen KL, Semon MJ et al. Sensing structural intermediates in bacterial flagellar assembly by export of a negative regulator. Science 1993; 262:1277-1280.

CHAPTER 7

SIGNAL TRANSDUCTION AND VIRULENCE GENE REGULATION IN *SHIGELLA* SPP.: TEMPERATURE AND (MAYBE) A WHOLE LOT MORE

Catherine M. C. O'Connell, Robin C. Sandlin
and Anthony T. Maurelli

1. INTRODUCTION

Many bacterial pathogens of man pass through several different environments before coming into contact with and colonizing the host. These may include another mammal (zoonotic infection), insects (arthropod-borne infections), or specific niches in the terrestrial or aquatic ecosystems, e.g. contaminated food or water. Even after gaining access to a host, a pathogen often passes through several different environments before reaching the site it will ultimately colonize. Thus, for most bacterial pathogens one can imagine a life cycle which consists of a segment outside the host and a segment within the host, the latter which can be subdivided into the various anatomical sites through which the organism passes on its way to the site of colonization.

Bacteria economize energy by regulating gene expression such that genes are turned on only when they are needed. Virulence genes are no exception and the various systems in the pathogens described in

The opinions or assertions contained herein are the private ones of the authors and are not to be construed as official or reflecting the views of the Department of Defense or the Uniformed Services University of the Health Sciences.

Signal Transduction and Bacterial Virulence, edited by Rino Rappuoli,
Vincenzo Scarlato and Beatrice Aricò. © 1995 R.G. Landes Company.

this book attest to this fact. For the most part, virulence genes are nonessential for survival outside the host. Expression of some virulence genes may even be detrimental to the organism when expressed outside the host. At the very least, such expression would be a waste of valuable energy. Therefore, bacterial pathogens have evolved to tightly control expression of their virulence genes. To express these genes only when in the host, bacteria need to be able to know when they have arrived in the host. The ability to sense signals which indicate to a bacterium that it is in a host must then be integrated into a regulatory loop such that the virulence genes can be turned on. Thus, signal sensing and signal transduction within the host is a critical element for success as a pathogen.

The organism which we study, *Shigella flexneri*, has proven to be an excellent model for the study of virulence gene regulation through environmental sensing. Bacteria of the species *Shigella* (*S. dysenteriae, S. flexneri, S. boydii*, and *S. sonnei*) are Gram-negative rods which are the causative agents of bacillary dysentery or shigellosis. Clinical disease is a result of invasion of the bacteria into the epithelial cells of the large intestine, multiplication within the cytoplasm of these cells and subsequent death of the infected cells. In the course of the infection, the bacteria spread laterally into adjacent cells in the epithelium, thus propagating the infection. These steps in the life cycle of *Shigella* in the human host can be directly correlated with the clinical symptoms of shigellosis: fever, severe abdominal cramps, diarrhea, and the passage of bloody, mucoid stools.[1]

Virulence of *Shigella* species is a multigenic phenomenon. Genes essential for virulence have been identified on both the chromosome and a large 180-220 kilobase virulence plasmid (for a review, see refs. 2 and 3). The large number of unlinked loci required for virulence imposes a requirement for coordinate regulation. This need and the need for temporal control

(within the host) has been met by linking the virulence genes in a regulon which responds to specific environmental signals. In this chapter we will review two of the best documented of these signals, temperature and osmolarity and we will describe what is known about the effectors of signal transduction. In addition, we speculate on other environmental signals which *Shigella* encounter on its passage through the gastrointestinal tract and how these might act on virulence gene regulation.

2. ENVIRONMENTAL SIGNALS, TARGETS, AND GENE EXPRESSION

2.1. TEMPERATURE

One very strong environmental cue which could be exploited by human pathogens to regulate virulence genes is temperature. The temperature within the human host is generally 37°C (slightly lower at the extremities), while outside of the host the temperature is usually less than 37°C. A shift from 25°C to 37°C could signal to the bacteria that it had arrived in the host and virulence genes should now be turned on. *Shigella* uses just such a change in temperature as a signal to regulate the expression of its virulence phenotype. Wild-type strains are fully virulent and invade cultured mammalian cells when grown at 37°C. When the same strains are grown at 30°C, they become phenotypically avirulent and cannot invade mammalian cells. Shifting the growth temperature back to 37°C restores virulence to the strains and de novo protein synthesis at 37°C is required for reexpression of the virulent phenotype.[4] A list of the virulence-associated phenotypes of *Shigella* which are regulated by growth temperature is given in Table 7.1.

Shigella is not unique in employing temperature as a signal for regulation since many other bacterial pathogens also regulate virulence gene expression in response to temperature.[5] Two of these bacterial pathogens, *Listeria monocytogenes* and *Yersinia enterocolitica* are the subjects of

Table 7.1. Virulence-associated phenotypes of Shigella spp. regulated by growth temperature

Invasion of mammalian cells
Intracellular multiplication (plaque assay)
Keratoconjunctivitis production (Sereny test)
Contact hemolytic activity
Pigmentation on Congo red agar
Shiga toxin production (*S. dysenteriae* 1)
Expression of the positive regulator *virB*
Expression of virulence genes:

ipaBCD	invasion plasmid antigens
mxi / spa operons	secretion of Ipa products
icsA; icsB	actin polymerization and cell-to-cell spread

separate chapters in this volume. Thus, the molecular mechanism of temperature-regulated virulence gene expression in *Shigella* may serve as a model system of gene regulation in response to temperature.

Operon fusion technology[6] has proven to be a powerful tool for studying gene regulation and we have applied this technique to the problem of temperature regulation of *Shigella* virulence genes. A suicide vector carrying a transposon with a promoterless "reporter" gene such as *lacZ* is used to generate random insertions into any bacterial genome. Fusions of the reporter gene to an active promoter place the reporter under the transcriptional control of the target gene promoter. Thus, even if the product of a gene is unknown or difficult to measure, the activity of its promoter can be determined by measuring the product of the fused "reporter" gene. In the case of the *lacZ* reporter, which encodes the enzyme β-galactosidase, promoter strength and activity can easily be measured on indicator media or in a quantitative enzyme assay.

We employed the *lacZ* reporter system to create operon fusions to temperature-regulated promoters in *S. flexneri* 2a with the aim of identifying virulence gene promoters.[7,8] Random insertions of a promoterless *lacZ* gene were generated followed by selection and screening for fusions to temperature-regulated promoters

(i.e. Lac⁺ at 37°C and Lac⁻ at 30°C). The fusions were then screened for loss of one or more virulence-associated phenotypes. This strategy resulted in the isolation of fusions to previously identified virulence genes but also allowed us to identify novel virulence genes.[8,9] Thus the operon fusion approach to studying virulence gene regulation by temperature yields the added bonus of providing an indirect means of identifying new virulence genes based on their response to this signal.

As the number of genes involved in *Shigella* pathogenesis grows, so does the number of temperature-regulated genes. A list of the known temperature-regulated virulence genes in *S. flexneri* is shown in Table 7.2. These genes include both structural as well as regulatory genes. It is worth noting that the ratio of gene expression at 37°C vs. 30°C shows a dramatic induction at the higher temperature. These levels underline the strength and importance of temperature as a signal since expression of the virulence genes can be induced to high levels from essentially an "OFF" position upon shift in temperature to 37°C.

Little is known about the kinetics of induction of virulence gene expression after a shift from 30°C to 37°C. However, preliminary studies in our laboratory have shown that there is a lag of about 30-40 minutes after temperature shift before expression of the structural virulence genes

Table 7.2. Thermo-induction of Shigella flexneri *virulence genes as measured by* lacZ *operons fusions and Northern blots (virF and virB)*

Strain	Fusion	β-galactosidase		Ratio 37/30°C	Ref.
		37°C	30°C		
BS228	*ipaB::lacZ*	382	6.3	60	8
BS232	*mxiA::lacZ*	468	4.4	106	8
BS226	*mxiB::lacZ*	550	11	50	8
BS230	*mxi17.7::lacZ*	668	14	49	8
BS184	*mxiC::lacZ*	167	6.5	25.7	32
SF403	*mxiD::lacZ*	620	30	21	83
SF132	*icsB::lacZ*	303	19	16	20
YSH6000	*virF*	Northern analysis		4	10
YSH6000	*virB*	Northern analysis		20	10

can be detected (Rowley et al, unpublished data). These experiments were performed using *lacZ* operon fusions to *ipaB*, *mxiA*, and *mxiB*, virulence genes which are in independent transcription units. The results suggest that expression of positive transcriptional activators may precede expression of the structural virulence genes. Indeed, when the induction of expression of the transcriptional activators *virF* and *virB* was measured after a temperature shift, it was found that expression of *virF* was almost immediate while induction of *virB* began about 20 minutes later. These results are consistent with the proposed model of VirF as an activator of *virB* expression and that of VirB as the direct mediator of expression of structural virulence genes such as *ipaB* and *mxiA*.[10] The role of these positive regulatory genes as well as a putative repressor gene (*virR*) as signal transducers of the temperature response will be addressed in a later section.

2.2. OSMOLARITY

Given the various environmental stimuli that shigellae encounter on their passage through the intestinal tract it would be surprising if the bacteria didn't use more than one of these as a signal for gene expression. Although temperature is clearly the strongest signal for *Shigella* virulence gene expression, osmolarity has

been shown to play a role as well and in this regard, *Shigella* are like several other human pathogens. Expression of alginate, the exopolysaccharide slime produced by *Pseudomonas aeruginosa*, responds to high osmolarity.[11,12] (See also chapter 3.) Several virulence determinants of *Vibrio cholerae*, including cholera toxin and toxin coregulated pili, are regulated by osmolarity.[13] (See also chapter 5.) The ability of *Salmonella typhi* to adhere to and invade human epithelial cells is also subject to modulation by osmolarity as is expression of the invasion gene *invA* of *S. typhimurium*.[14,15]

The *ompB* locus of *Escherichia coli* K-12 (map position 75 min. on the chromosome) encodes a two component system which regulates outer membrane porin expression in response to changes in osmolarity.[16] The *S. flexneri* homologue of *ompB* not only regulates porin expression but also plays a role in virulence gene expression thus suggesting that osmolarity might be another environmental signal for *Shigella*.[17] Mutations in *ompB* reduce the ability of *S. flexneri* to invade into and multiply within mammalian cells. The defect in virulence is not due to altered expression of the Ipa proteins (invasion plasmid antigens) which are required for invasion since levels of Ipa synthesis are unaltered in *ompB* mutants. However, the outer membrane porins

OmpF and OmpC are not expressed in an *ompB* mutant and recent data suggest that *ompB* mutants may be attenuated for virulence because they fail to synthesize OmpC. Thus, mutants which are defective only in expression of OmpF retain virulence while mutants which synthesize OmpF but not OmpC are severely impaired in their virulence primarily at the level of cell to cell spread and host cell killing.[18] The absence of this major porin molecule from the outer membrane of *S. flexneri* can have a significant impact on expression of other outer membrane proteins including possibly virulence determinants. At present, it is not clear whether OmpC plays a role in virulence via its function as a porin or indirectly via its interactions with other surface membrane molecules.

Interestingly, data that call into question the role of osmolarity as a signal for virulence gene expression can be cited. Unlike the case with *E. coli* K-12, where expression of *ompC* is regulated by the *ompB* locus through changes in osmolarity, expression of *ompC* in *S. flexneri* is dependent on *ompB* but expression is constitutive at high and low osmolarity.[18] This observation should be contrasted with data which show that at least two virulence genes of *S. flexneri* (*mxiC* and *icsB*) are regulated by changes in medium osmolarity. Quantitative measurement of *icsB* mRNA and expression from an operon fusion to *mxiC* demonstrate that both virulence genes are optimally expressed at high osmolarity at 37°C.[18,19] However, the levels of induction from low to high osmolarity only vary by 3- to 7-fold, suggesting that this form of regulation is secondary to regulation by temperature. It is important to note that *icsB* is a gene whose product is implicated in cell-to-cell spread of the bacteria after escape into the host cell cytoplasm.[20] Therefore, the need for expression of *icsB* would come after the bacteria have invaded the mammalian cell. This step represents a transition from the low osmolarity environment of the intestinal lumen into the high osmolarity environment of the mammalian cell cytoplasm. So it is reasonable to postulate that osmolarity could act as a signal for expression of at least a subset of virulence genes of *Shigella*, perhaps those genes which are required during the intracellular stage of the pathogen's life cycle. The subject of the intracytoplasmic environment as a signal for *Shigella* virulence gene expression will be addressed below.

2.3. OTHER POTENTIAL ENVIRONMENTAL SIGNALS

2.3.1. pH

Several bacteria have pH-regulated genes including *S. typhimurium*, *E. coli*, and *V. cholerae*.[21-23] The *atp* operon (encoding a Mg^{2+}-dependent proton-translocating ATPase) and the *fur* gene (encoding a Fe^{2+}-binding regulatory protein) in *S. typhimurium* are involved in acid tolerance and are induced in mildly acidic conditions. In addition to being extremely acid sensitive, *atp* mutants are avirulent in the mouse typhoid model by oral and intraperitoneal routes.[24]

During transition to the site of invasion in the large intestine, *Shigella* are exposed to extremes of pH moving from very low pH in the stomach to high pH in the small intestine. *Shigella* have the ability to resist the low pH of the stomach as the infectious dose 50% (ID_{50}) is less than 500 organisms.[25] A comparison of survival times of *Salmonella*, *Shigella*, and enteroinvasive *E. coli* at different pH demonstrated that only *S. flexneri* is able to survive at pH 2.[26] These results indicate that the ability to survive in the low pH environment of the stomach correlates with low ID_{50}. A gene involved in acid resistance in *S. flexneri* has been cloned and shown to be the *Shigella* homologue of *rpoS*/*katF* which encodes the growth phase-dependent sigma factor σ^{38} in *E. coli* K-12. This gene is required for extreme acid resistance in aerobic cultures of *S. flexneri* but the degree of dependence is influenced by growth phase and/or oxygen levels.[27]

Following invasion of epithelial cells, *Shigella* are again exposed to low pH in the phagocytic vacuole. Upon lysis of the phagocytic vesicle and release into the cytoplasm, they return to a neutral pH environment. Growth of *Shigella* in minimal medium with ion concentrations and pH similar to that of the intracellular compartment results in the induction of 97 kDa and 58 kDa proteins.[28] The genes encoding the 97 and 58 kDa proteins have not yet been identified. Nevertheless, this result suggests that pH, and perhaps other intracellular signals, can regulate gene expression in *Shigella*.

2.3.2. Intracytoplasmic Environment

As noted above, the environment within a eukaryotic cell is very different from the extracellular environment of the intestinal lumen. Following invasion, *Shigella* encounter a pH drop in the endocytic vacuole and then a return to neutral pH following escape to the cytoplasm. In the extracellular body fluids, sodium and calcium ion concentrations are high while magnesium and potassium levels are low. However these conditions are reversed in the cytoplasm where magnesium and potassium levels rise and sodium and calcium ion concentrations decrease.[28] The cytoplasm is a reducing rather than an oxidizing environment. The drastic change in environmental conditions following entry into the host cell could induce a change in the bacterial proteins synthesized. Radiolabeling with [^{35}S]methionine under different growth conditions was used to demonstrate altered protein profiles in *S. flexneri*.[28] Labeling of HeLa cell monolayers infected with *Shigella* demonstrated that both repression and upregulation of several proteins occurs in vivo in addition to the appearance of novel proteins that are only detected following in vivo growth. Time course studies correlate specific induced/upregulated proteins with different stages of infection. The expression of some of the induced proteins can be detected in vitro following growth of *S. flexneri* in minimal media formulated

to mimic the pH, reducing environment, and ion concentrations of the intracellular compartment. Immunoprecipitation studies using monkey convalescent serum identified proteins from in vivo conditions that are not detected during in vitro conditions.

The results of these studies demonstrate that altered expression of proteins occurs during growth in tissue culture monolayers and suggests that differential protein expression in *Shigella* occurs in vivo. However, more sophisticated approaches to studying in vivo gene expression need to be applied. New techniques for studying in vivo expression have resulted in the identification of potential virulence genes in other pathogens. For example, expression of the *spvB* gene of *Salmonella dublin* is required for mouse virulence and this gene is rapidly induced following entry into macrophages, epithelial cells, and hepatocytes.[29] Several genes in *S. typhimurium* are induced or repressed following entry into the macrophage and these are regulated by a two-component regulatory system, PhoP/PhoQ.[30] (See also chapter 4.) A genetic system for the identification of promoters that are activated in vivo has been devised and has identified several novel loci in *S. typhimurium* that are upregulated in vivo.[31] This system is ideally suited for pathogens for which in vivo selection models are available. Although originally designed for use with *Salmonella* in mouse infections, the system can be adapted to identify genes in *Shigella* that are specifically induced during replication in tissue culture cells. The adaptation of these new techniques to the study of virulence factors of *Shigella* will more clearly answer the question of whether intracellular signals regulate virulence gene expression and help identify what the transducers of those signals might be.

3. SIGNAL TRANSDUCERS

3.1. VirR/H-NS

In order to identify elements involved in temperature regulation of *S. flexneri* virulence genes, a strain carrying a *mxiC::lacZ* fusion was mutagenized using Tn*10* and mutants were selected which were constitutive for the Lac+ phenotype at 30°C, the nonpermissive temperature for expression of the virulence genes.[32] When these mutations are transduced into a wild type strain, all the transductants become derepressed for virulence and are capable of invading HeLa cells at both 30°C and 37°C. These transposon insertions map to the chromosome at 27 minutes and the locus is designated *virR* because of its ability to repress virulence gene expression in *S. flexneri*. Subsequent cloning and sequencing of this locus[33] showed it to be identical to *hns* which encodes a histone-like protein of 15 kDa initially identified as a major component of the *E. coli* nucleoid.[34,35] H-NS is a neutral protein (pI = 7.5), abundant in stationary phase cells and binds preferentially to double stranded DNA as a homodimer.[36] Homologues of this gene have been identified in other members of the family Enterobacteriaciae.[37,38] Genetic evidence for a role for this protein as a global repressor in prokaryotes is based on the mapping of mutant regulatory alleles such as *bglY*,[39] *drdX*[40] and *osmZ*,[41] to the *hns* locus. Genes such as *adi*, *cadA*,[42] the *pap* operon of uropathogenic *E. coli*[40] and the *proVWX* operon[41] are regulated by *hns* in response to diverse environmental stimuli, e.g. pH, temperature and osmolarity. In addition, it has been observed that the expression of many other cellular proteins is either strongly induced or repressed in a *hns* mutant.[43,44] Mutations in *hns* have other pleiotropic effects: decreasing bacterial motility,[43,45] increasing the frequency of chromosomal deletions,[39] modulating the inversion rate of the DNA fragment containing the promoter for type 1 pili[46] and stimulating the transposition of bacteriophage Mu.[47] In most instances studied, H-NS appears to act as a transcriptional repressor, one prominent exception being the regulation of the *ilvIH* operon where H-NS is believed to modulate activation by the leucine regulatory protein.[48]

In addition to regulating the virulence genes of *S. flexneri*, H-NS is an important regulator of virulence genes in *E. coli* and *S. typhimurium*. H-NS regulates expression of the *pap* locus in uropathogenic strains of *E. coli*[40] and the CFA/I fimbrial operon of enterotoxigenic *E. coli*.[49] These operons are both regulated in response to temperature. *hns* mutants of mouse virulent strains of *S. typhimurium* and *S. enteritidis* are attenuated and virulence is restored when genetic crosses reintroduce a wild type *hns* allele.[50]

The molecular mechanisms by which *hns* acts as a repressor are not fully understood although several hypotheses have been suggested. One model is that the protein acts as a transcriptional silencer which renders parts of the chromosome inactive by H-NS induced compaction of the nucleoid in a manner analogous to silencing in eukaryotic systems.[40,51] Some reservations about this hypothesis can be found in the observations that mutations in *hns* affect gene expression on a specific and localized basis and regulation is apparently instantaneous in response to environmental stimuli.[52] In fact, in vitro studies of the interaction of H-NS with *proU*, the promoter of the osmotically inducible *proVWX* operon, show that the inhibitory effect of H-NS on *proV* transcription is both specific and local, with no effect on an adjacent, H-NS insensitive promoter carried on the same DNA template.[53]

A second hypothesis postulates that transcriptional regulation by H-NS is due to alterations of DNA supercoiling mediated by binding of DNA by this protein.[52] Changes in supercoiling are rapid, do not appear to require protein synthesis and have been shown to affect transcription of several genes.[54] Experiments in *Salmonella* show that *hns* mutations result in increased amounts of supercoiling and that the increased supercoiling correlates with derepression of *proU* in the absence of osmotic shock.[45] However, H-NS can specifically inhibit transcription from *proU* in an in vitro system containing only

RNA polymerase and supercoiled DNA where no other components which could alter the topology of the DNA are present.[53] Spassky et al reported that binding of linear or circular DNA by H-NS has little effect on the superhelicity of the molecule.[36] Identical *virR*::Tn*10* mutations in *E. coli* and *S. flexneri* exhibit opposite effects in vivo on reporter plasmid supercoiling although the cloned *E. coli* locus regulates virulence gene expression in *S. flexneri* in a manner identical to the *S. flexneri virR* gene.[55,56] A better understanding of the topology of DNA in *hns*-sensitive promoters when exposed to appropriate environmental stimuli is required in order to determine the contribution that DNA supercoiling may make to repression mediated by H-NS. A third model suggests that H-NS acts in a manner similar to classical repressors by selectively binding to the promoter of target genes. This binding could occur through the recognition of a consensus sequence. Analysis of H-NS binding to the *gal* and *lac* promoters led to a putative consensus sequence 5'-TNTNAN-3'.[57] However, results from several studies suggest that H-NS recognizes a specific DNA structure such as curved DNA.[43,58,59] In support of this hypothesis, Yamada et al[43] demonstrated that H-NS selectively binds to curved DNA in vitro and several groups have shown that genes regulated by H-NS, including *proU*[59,60] and *hns* itself,[61,62] have curved DNA in upstream regions. Expression of *hns* is autoregulated and takes place at the transcriptional level.[61,62] DNA footprinting analysis of the promoter region of the *hns* promoter indicates that H-NS binds three sites, including a site of relatively low affinity which spans the -35 region of the promoter. This site is bound after the other two and it is envisaged that autorepression requires the cooperative binding of several H-NS dimers which would ultimately occlude access of RNA polymerase to the -35 box.[61] Multiple binding sites for H-NS have also been identified for *proU*.[60] The finding that a mutant H-NS protein with near-normal DNA binding capacity but which is defective in tetramer formation is unable to recognize the DNA curve 5' to *hns* and does not repress transcription of the gene, lends support to this hypothesis.[61]

In *S. flexneri*, repression by VirR/H-NS occurs at the level of transcriptional regulation of another activator, *virB*. DNA footprinting experiments show that H-NS can specifically bind *virB* DNA, in a single large footprint, in the region spanning from +20 to -20 which includes the transcriptional start site of *virB* and the -10 region.[63] Binding by H-NS to this site could presumably block binding of the RNA polymerase to the *virB* promoter. S1 nuclease protection assays of transcripts from plasmids carrying deletions upstream of the *virB* promoter confirmed that this region is essential for in vivo repression by H-NS. A sequence in this region between +3 and -3 (TGTGAG) was identified which matches the consensus for H-NS binding described above. However, other potential H-NS consensus sequences exist outside this region. Preliminary data suggest that the DNA in the region +173 through to -110 is curved. Interestingly, while H-NS was demonstrated to specifically bind DNA encompassing the *virB* promoter even in the presence of competitor DNA, only one H-NS footprint was identified, unlike the multiple binding sites identified around the promoters of the *proV* operon and *hns*. As yet, no other genes in the *Shigella* virulence regulon have been shown to have any interaction with H-NS and it is presumed that the signal for temperature induction is somehow transduced via H-NS interaction with the *virB* promoter.

How a temperature signal, or indeed any other environmental stimulus, could be transduced via H-NS is unclear. Expression from *hns* is stimulated 4-fold by the transcriptional enhancer encoded by *cspA* in response to a temperature shift from 37°C to 10°C.[64] The amount of H-NS present in the cell correlates with the growth phase of the bacterial culture, increasing 10-fold from early exponential to stationary phase and this reflects a 10-fold increase in expression from *hns*.[62] Competitive binding of the *hns* promoter by Fis, another histone-like protein, serves to alleviate some autorepression in early-log growth.[61] Thus the amount of H-NS present in the cell is tightly regulated at all times and derepression of the *Shigella* virulence loci in response to temperature presumably does not reflect a reduction in the amount of H-NS in the cytoplasm. In a model of temperature regulation suggested for *Y. enterocolitica* it is envisaged that curved DNA bound by a histone-like protein melts at 37°C and dislodges the repressor protein leaving the site available for binding by a transcriptional activator.[65] Alterations in DNA supercoiling in response to an environmental stimulus such as temperature have been envisaged to have similar effects.[52] This model could have implications for *Shigella* which requires the trancriptional activator VirF for expression of the other virulence genes (see below).

Hromockyj et al proposed an alternative model suggesting that oligomerization of VirR/H-NS is required for the recognition of specific binding sites.[33] They hypothesized that oligomerization occurs in a concentration-dependent manner since in vitro experiments show that H-NS can form trimers and tetramers at high concentrations.[66] There is no evidence that *hns* is more highly expressed at 30°C than at 37°C. Nonetheless, the observation that H-NS cooperatively binds specific sites in sensitive promoters in multimeric forms suggests that there is merit in some aspects of this model. The conditions which favor the formation of H-NS multimers in vivo are not known but any mechanism which would destabilize binding of H-NS to DNA could result in derepression. Dissolution of H-NS/DNA complexes could result from direct, temperature-induced conformational changes disrupting protein-protein interactions or through loss of an essential co-factor or protein required to maintain the H-NS complexes. A second site mutation which suppresses a *hns* mutant phenotype has been isolated in

hscA, a Hsp70 protein believed to have chaperone-like properties which could ultimately influence the conformation of H-NS.[67] Different isoforms of H-NS could modulate the extent of multimerization or the stability of H-NS/DNA complexes. Thus, regulation of sensitive operons would take place at a purely translational level. Bacteriophage T7 has evolved a protein, gp5.5, which binds to a DNA/H-NS complex in vitro and abolishes H-NS mediated inhibition of transcription by *E. coli* RNA polymerase.[68] This interaction presumably prevents repression by the bacterial H-NS of bacteriophage genes required for replication. Homologues of this T7 protein may exist in *Shigella* and could be induced in response to an environmental signal and in this way result in derepression. The diverse nature of H-NS sensitive operons and the wide range of signals and environmental conditions under which regulation occurs suggest that components which add specificity to the regulation of individual systems are essential. No single, universal mechanism may account for the modulation of H-NS repression of these diverse operons.

3.2. VirF

The *virF* locus of *S. flexneri* encodes a 30 kDa protein which positively regulates expression of the virulence genes *ipaBCD* and *virG*.[69,70] *virF* also positively regulates the expression of *virB*.[71] Comparison of the amino acid sequence of VirF with sequences contained within the EMBL and GenBank data bases has revealed that the protein contains a helix-turn-helix motif in the C-terminal portion of the protein which is characteristic of the AraC family of transcriptional activators.[72] Many examples of AraC-like regulators have been identified and estimations of the evolutionary relatedness of the various family members have been attempted.[73] An example of such an alignment is shown in Figure 7.1. This dendrogram was constructed using the CLUSTAL multiple alignment program[74] and while it cannot give any estimate of evolutionary distance, a good estimate of branching order may be made. The AraC family has two major clusters representing regulators of metabolite utilization pathways (AraC, MelR) and proteins which are involved in the regulation of genes in response to a variety of environmental stimuli e.g. temperature, growth phase. Examples of members of the latter group include the positive activators CfaD and Rns, both of which regulate pilus expression in response to temperature.[75,76]

The proteins of the VirF sub-family have in common a much higher degree of homology which extends over the entire protein. These homologous regions may include domains involved in function or sensing common to this group. The VirF homologues are basic (pI ~ 9), about 30 kDa in size, and are often plasmid encoded. The genes encoding these pro-teins are often very A+T rich. Some of the operons regulated by these genes are also repressed by H-NS. Finally, most are involved in the thermoregulation of virulence genes. This suggests a common ancestor for these regulators and therefore some degree of commonality in terms of their role in regulating virulence genes and in how they themselves are regulated. Striking exceptions to this group are the VirF and LcrF genes of *Yersinia* spp. A review of the role these proteins play in regulating *Yersinia* virulence gene expression is presented elsewhere in this volume, however, it should be noted that although VirF and LcrF are encoded on the virulence-associated plasmid present in these organisms, and activate the expression of virulence genes in response to temperature, they are more homologous to AraC-like proteins than to the VirF sub-family.

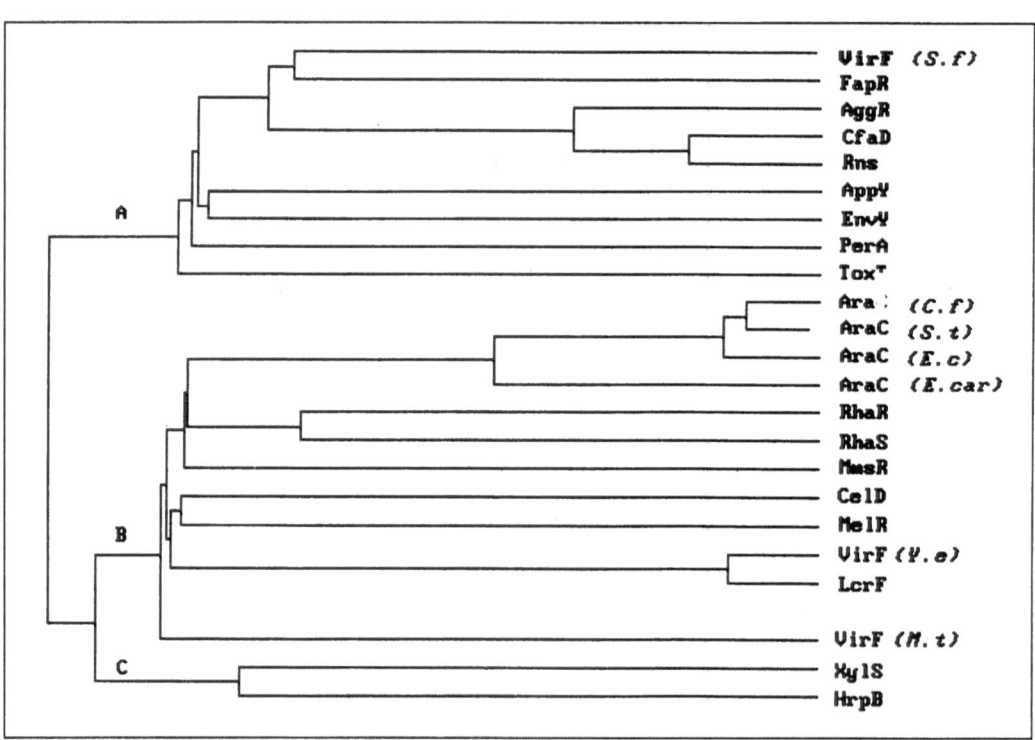

Fig. 7.1. AraC family of transcriptional activators. Dendrogram of relatedness between family members was generated using the CLUSTAL multiple alignment program. Amino acid sequences of family members were obtained from GenBank Release 84.0 (8/94) with the exception of PerA (Kaper J, personal communication). Bacterial species are abbreviated as follows: S. f, Shigella flexneri; C. f, Citrobacter freundii; S. t, Salmonella typhimurium; E. c, Escherichia coli; E. car, Erwinia caratovara; Y. e, Yersinia enterocolitica; M. t, Mycobacterium tuberculosis.

The mechanism by which VirF activates transcription of *virB* is not understood. DNA footprinting experiments using a VirF fusion protein indicate that VirF binds target sequences in the upstream region of *virB*.[77] It is possible that VirF acts antagonistically as an antirepressor, competing with H-NS for separate but overlapping sites in the *virB* promoter. A similar model has been proposed for the regulation of CFA/I fimbriae by CfaD, a VirF homologue.[49] Interestingly in the temperature-regulated virulence operons of other pathogens which contain *virF* homologues, a *virB* homologue is absent. Perhaps the promoters of the structural genes in these operons have 5' regulatory regions which resemble that of *virB*, with both H-NS and "VirF" binding sites and shigellae apparently evolved VirB because of the large number of virulence genes which must be coordinately regulated.

As discussed above, the mechanism by which a temperature signal is transduced to the *Shigella* virulence regulon is not clear and it is not known whether VirF plays any role in sensing a temperature shift. The *virF* gene itself is subject to weak temperature regulation, its transcription increasing only 4-fold at 37°C compared with levels at 30°C.[71] Additionally it has been reported that increasing the amount of H-NS in the cell does not correlate with enhanced transcriptional repression of *virF* from enteroinvasive *E. coli*.[78] This suggests that any temperature sensing by VirF may take place at the translational level. In vivo a shift from 30°C to 37°C might enhance the DNA-binding ability of the protein and allow it to out-compete H-NS for the appropriate target sequences.

3.3. VIRB

Activation of the *ipa* operon and the *mxi/spa* operons by VirF is mediated by a second transcriptional activator, VirB. VirB was originally identified as a 35.4 kDa protein encoded on the virulence plasmid of *S. flexneri* 2a[71] and functional homo-

logues of *virB*, which are almost identical at the DNA sequence level, have been found in *S. flexneri* 5 (*ipaR*)[79] and *S. sonnei* (*invE*).[80] The calculated pI of VirB (pI = 9.7) indicates a basic, positively charged protein and its hydrophobicity profile demonstrates several hydrophilic regions characteristic of cytosolic proteins.[79]

The expression of VirB itself is dependent on both activation by VirF and temperature.[10] Expression of *virB* is completely abolished in a *virF* mutant. As for the temperature requirement, *virB* is "OFF" at 30°C and "ON" at 37°C. Overexpression of VirF at 30°C can restore transcription of *virB* but not to the level seen at 37°C. The temperature requirement for transcription of the virulence genes can be bypassed by overproduction of VirB at 30°C. These results demonstrate that expression of *virB* is absolutely dependent on VirF but temperature provides an additional essential signal for transcription.

To define the regulatory regions involved in *virB* expression, progressive 5' deletions of the DNA upstream of the transcription start point of *virB* were constructed and placed upstream of the chloramphenicol acetyltransferase gene to monitor promoter activity in the presence and absence of VirF. The results indicate that VirF may act with varying affinities at two sites upstream of the *virB* start site, one site downstream of -59 and the second site upstream of -104 (relative to the start codon of *virB*).[81] These results were independently confirmed by using S1 nuclease protection assays to monitor promoter activity of *virB* promoter deletion constructs.[77]

VirF can also activate *virB* transcription when it is used as a fusion protein. A fusion protein was constructed by fusing the 3' portion of *malE* to the *virF* coding sequence lacking the first 10 base pairs.[77] When a crude lysate of a strain expressing the MalE-VirF fusion protein was purified on an affinity column made with the *virB* promoter region, two proteins of 70 kDa and 16 kDa were purified. The 70 kDa protein was shown to be MalE-VirF and the 16 kDa species was shown to be H-NS. DNA footprinting analysis demonstrated that H-NS bound to a region from +20 to -20 and therefore covered the -10 sequence and the transcription start site of *virB*. Due to the instability of VirF, footprinting experiments were conducted with the fusion protein. The footprint of *malE-virF* fusion protein covered a region from -17 through -105 and blocked the -35 region. Considering the size of the VirF footprint and the results of the promoter deletion analysis, VirF may be binding at two sites in the *virB* promoter region.[77] These results suggest that H-NS binding directly blocks transcription of *virB* by blocking part of the RNA polymerase binding site and the *virB* start codon. However, the mechanism by which VirF and temperature overcome this block is unknown.

Sequence analysis of *virB* (*ipaR/invE*) demonstrates that this protein does not contain motifs associated with known transcriptional regulators including TdcR, Fis, NtrC, TnpR, Hin, and the LysR family.[79,80] VirB also does not contain the Ser-Pro-X-X or Thr-Pro-X-X motif that often distinguishes regulatory proteins from nonspecific DNA binding proteins. However, significant amino acid identity exists between VirB and the bacteriophage P1 ParB protein and the SopB protein of F plasmid.[79] ParB is a site-specific DNA binding protein that recognizes a 13 base pair inverted repeat half-site in the 34 base pair *parS* locus of the P1 plasmid.[82] The ParB/*parS* complex interacts with ParA to effect partitioning of the replicated P1 molecules into the daughter cells. Despite the sequence homology with ParB and SopB, neither the *sopB* gene nor the *parB* gene can complement the function of *virB*.[80]

Very little is known about the mode of VirB activation of the *ipa* genes, *mxi* genes, and *spa* genes. No potential consensus *virB* binding sites have been identified in the promoter regions of the *ipa* and *mxi/spa* operons.

4. PERSPECTIVES

Regulation of virulence gene expression in *Shigella* spp. presents a complex set of problems for the bacteria. Virulence genes are encoded on a plasmid and on the chromosome. Ideally, these genes should only be expressed once the bacterium has entered its human host. Further, some genes are needed early in infection while others are needed after the bacterium has moved from the intestinal lumen into the cytoplasm of the epithelial cell. Coordinate regulation of the virulence regulon in response to temperature is the solution that has evolved in *Shigella* as well as in other bacterial pathogens of man. While temperature is the major signal for induction of virulence gene expression in *Shigella*, other environmental stimuli act as subtle modulators of expression. This "fine-tuning" expression of the already activated virulence regulon may have evolved to take advantage of the changing environments that the bacterium passes through. In this regard, the study of bacterial gene expression inside the mammalian cell presents an exciting new area of research.

The %G+C of the virulence genes on the plasmid is in the range of 33-37% while the %G+C of the chromosome of *Shigella* is much higher at 50%.[2] Since the plasmid-encoded virulence genes are subject to control by chromosomal regulators such as *virR/hns* and *ompB*, it is remarkable that these "foreign" genes evolved or adapted to place themselves under the control of chromosomal genes. A more perplexing mystery is how VirR/H-NS plays the pivotal role in temperature regulation in *Shigella* while being an important temperature-independent repressor for other genes in other enteric bacteria.

A better understanding of the precise role of VirR/H-NS in repressing virulence gene expression holds out the prospect of developing new therapeutic approaches to treating *Shigella* infection. Similarly, understanding how VirF and VirB act to upregulate expression of their respective target genes may allow development of agents which can interfere with their activity. Therefore pursuit of these fundamental questions of signal transduction and gene regulation in *Shigella* as well as other bacterial pathogens will serve not only to broaden our knowledge of bacterial pathogenesis but has the potential to lead to immediate application in the area of prevention and treatment of disease.

ACKNOWLEDGMENTS

Research in our laboratory is supported by grant AI24656 (to A.T.M.) from the National Institute of Allergy and Infectious Diseases and USUHS protocol RO-7385. R.C.S. is supported by training grant T32 AI07308 from the National Institutes of Health.

REFERENCES

1. Sansonetti PJ. Molecular and cellular biology of *Shigella* invasiveness: From cell assay systems to shigellosis. In: Sansonetti PJ, ed. Pathogenesis of Shigellosis. Berlin: Springer-Verlag, 1992: 1-19.

2. Hale TL. Genetic basis of virulence in *Shigella* species. Microbiol Rev 1991; 55:206-224.

3. Sasakawa C, Buysse JM, Watanabe H. The large virulence plasmid of *Shigella*. In: Sansonetti PJ, ed. Pathogenesis of Shigellosis. Berlin: Springer-Verlag, 1992:21-44.

4. Maurelli AT, Blackmon B, Curtiss III R. Temperature-dependent expression of virulence genes in *Shigella* species. Infect Immun 1984; 43:195-201.

5. Maurelli AT. Temperature regulation of virulence genes in pathogenic bacteria: a global strategy for human pathogens. Microbial Pathogen 1989; 7:1-10.

6. Silhavy TJ, Beckwith JR. Use of *lac* fusions for the study of biological problems. Microbiol Rev 1985; 45:398-418.

7. Maurelli AT, Curtiss R III. Bacteriophage Mu *d*1(Apʳ *lac*) generates *vir-lac* operon fusions in *Shigella flexneri* 2a. Infect Immun 1984; 45:642-648.

8. Hromockyj AE, Maurelli AT. Identification of *Shigella* invasion genes by isola-

tion of temperature-regulated *inv::lacZ* operon fusions. Infect Immun 1989; 57:2963-70.

9. Andrews GP, Hromockyj AE, Coker C et al. Two novel virulence loci in *Shigella flexneri* 2a, *mxiA* and *mxiB*, facilitate excretion of invasion plasmid antigens. Infect Immun 1991; 59:1997-2005.

10. Tobe T, Nagai S, Okada N et al. Temperature-regulated expression of invasion genes in *Shigella flexneri* is controlled through the transcriptional activation of the *virB* gene on the large plasmid. Mol Microbiol 1991; 5:887-893.

11. Deretic V, Dikshit R, Konyecsni WM et al. The *algR* gene, which regulates mucoidy in *Pseudomonas aeruginosa*, belongs to a class of environmentally responsive genes. J Bacteriol 1989; 171:1278-1283.

12. Berry A, DeVault JD, Chakrabarty AM. High osmolarity is a signal for enhanced *algD* transcription in mucoid and nonmucoid *Pseudomonas aeruginosa* strains. J Bacteriol 1989; 171:2312-2317.

13. Miller VL, Mekalanos JJ. A novel suicide vector and its use in construction of insertion mutations: osmoregulation of outer membrane proteins and virulence determinants in *Vibrio cholerae* requires *toxR*. J Bacteriol 1988; 170:2575-2583.

14. Tartera C, Metcalf ES. Osmolarity and growth phase overlap in regulation of *Salmonella typhi* adherence to and invasion of human intestinal cells. Infect Immun 1993; 61:3084-3089.

15. Galan JE, Curtiss III R. Expression of *Salmonella typhimurium* genes required for invasion is regulated by changes in DNA supercoiling. Infect Immun 1990; 58:1879-1885.

16. Ronson CW, Nixon BT, Ausubel FM. Conserved domains in bacterial regulatory proteins that respond to environmental stimuli. Cell 1987; 49:579-581.

17. Bernardini ML, Fontaine A, Sansonetti PJ. The two-component regulatory system OmpR-EnvZ controls the virulence of *Shigella flexneri*. J Bacteriol 1990; 172:6274-6281.

18. Bernardini ML, Sanna MG, Fontaine A et al. OmpC is involved in invasion of

epithelial cells by *Shigella flexneri*. Infect Immun 1993; 61:3625-3635.

19. Porter ME, Dorman CJ. A role for H-NS in the thermo-osmotic regulation of virulence gene expression in *Shigella flexneri*. J Bacteriol 1994; 176:4187-4191.

20. Allaoui A, Mounier J, Prevost M-C et al. *icsB*: a *Shigella flexneri* virulence gene necessary for the lysis of protrusions during intercellular spread. Mol Microbiol 1992; 6:1605-1616.

21. Aliabadi Z, Park YK, Slonczewski JL et al. Novel regulatory loci controlling oxygen- and pH-regulated gene expression in *Salmonella typhimurium*. J Bacteriol 1988; 170:842-851.

22. Parsot C, Mekalanos JJ. Expression of the *Vibrio cholerae* gene encoding aldehyde dehydrogenase is under control of ToxR, the cholera toxin transcriptional activator. J Bacteriol 1991; 173:2842-2851.

23. Watson N, Dunyak DS, Rosey EL et al. Identfication of elements involved in transcriptional regulation of the *Escherichia coli cad* operon by external pH. J Bacteriol 1992; 174:530-540.

24. Garcia-Del Portillo F, Foster JW, Finlay BB. Role of acid tolerance response genes in *Salmonella typhimurium* virulence. Infect Immun 1993; 61:4489-4492.

25. DuPont HL, Levine MM, Hornick RB et al. Inoculum size in shigellosis and implications for expected mode of transmission. J Infect Dis 1989; 159:1126-1128.

26. Gorden J, Small P. Acid resistance in enteric bacteria. Infect Immun 1993; 61:364-367.

27. Small P, Blakenhorn D, Welty D et al. Acid and base resistance in *Escherichia coli* and *Shigella flexneri*: role of *rpoS* and growth pH. J Bacteriol 1994; 176:1729-1737.

28. Headley VL, Payne SM. Differential protein expression by *Shigella flexneri* in intracellular and extracellular environments. Proc Natl Acad Sci USA 1990; 87:4179-4183.

29. Fierer J, Eskmann L, Fang F et al. Expression of the *Salmonella* virulence plasmid gene *spvB* in cultured macrophages and nonphagocytic cells. Infect Immun 1993; 61:5231-5236.

30. Miller SI. PhoP/PhoQ: macrophage-specific modulators of *Salmonella* virulence? Mol Microbiol 1990; 5:2073-2078.

31. Mahan MJ, Slauch JM, Mekalanos JJ. Selection of bacterial virulence genes that are specifically induced in host tissues. Science 1993; 259:686-688.

32. Maurelli AT, Sansonetti PJ. Identification of a chromosomal gene controlling temperature-regulated expression of *Shigella* virulence. Proc Natl Acad Sci USA 1988; 85:2820-2824.

33. Hromockyj AE, Tucker SC, Maurelli AT. Temperature regulation of *Shigella* virulence: identification of the repressor gene *virR*, an analogue of *hns* and partial complementation by tyrosyl transfer RNA (tRNA$_i^{Tyr}$). Mol Microbiol 1992; 6:2113-2124.

34. Pon CL, Calogero RA, Gualerzi CO. Identification, cloning, nucleotide sequence and chromosomal map location of *hns*, the structural gene for *Eshcerichia coli* DNA-binding protein H-NS. Mol Gen Genet 1988; 212:199-202.

35. Durrenberger M, La Teana A, Citro G et al. *Escherichia coli* DNA-binding protein H-NS is localized in the nucleoid. Res Microbiol 1991; 142:373-380.

36. Spassky A, Rimsky S, Garreau H et al. H1a, an *E. coli* DNA-binding protein which accumulates in stationary phase. Nucleic Acids Res 1984; 12:5321-5340.

37. Hulton CSJ, Seirafi A, Hinton JCD et al. Histone-like protein H1 (H-NS) DNA supercoiling and gene expression in bacteria. Cell 1990; 63:631-642.

38. La Teana A, Falconi M, Scarlato V et al. Characterization of the structural genes for the DNA-binding protein H-NS in Enterobacteriaceae. FEBS Lett 1989; 244:34-38.

39. Lejeune P, Danchin A. Mutations in *bglY* increase the frequency of spontaneous deletions in *Escherichia coli* K-12. Proc Natl Acad Sci USA 1990; 87:360-363.

40. Goransson M, Sonden B, Nilson P et al. Transcriptional silencing and thermo-

regulation of gene expression in *Escherichia coli*. Nature 1990; 344:682-685.

41. Higgins CF, Hinton JCD, Dorman CJ et al. A physical role for DNA supercoiling in the osmotic regulation of gene expression in *S. typhimurium* and *E. coli*. Cell 1988; 52:569-584.

42. Shi X, Waasdorp BC, Bennett GN. Modulation of acid-induced amino acid decarboxylase gene expression by *hns* in *Escherichia coli*. J Bacteriol 1993; 175:1182-1186.

43. Yamada H, Yoshida T, Tanaka K et al. Molecular analysis of the *Eshcerichia coli hns* gene encoding a DNA-binding protein, which preferentially recognizes curved DNA sequences. Mol Gen Genet 1991; 230:332-336.

44. Yoshida T, Ueguchi C, Yamada H et al. Function of the *Escherichia coli* nucleoid protein, H-NS: molecular analysis of a subset of proteins whose expression is enhanced in a *hns* deletion mutant. Mol Gen Genet 1993; 237:113-122.

45. Hinton JCD, Santos DS, Seirafi A et al. Expression and mutational analysis of the nucleoid-associated protein H-NS of *Salmonella typhimurium*. Mol Microbiol 1992; 6:2327-2337.

46. Kawula TH, Orndorff PE. Rapid site-specific DNA inversion in *Eshericia coli* mutants lacking the histone-like protein H-NS. J Bacteriol 1991; 173:4116-4123.

47. Kano Y, Yasuzawa K, Tanaka H et al. Propagation of phage Mu in IHF-deficient *Escherichia coli* in the absence of the H-NS histone-like protein. Gene 1993; 126:93-97.

48. Levinthal M, Lejeune P, Danchin A. The H-NS protein modulates the activation of the *ilvIH* operon of *Escherichia coli* K-12 by Lrp, the leucine regulatory protein. Mol Gen Genet 1994; 242:736-743.

49. Jordi BJAM, Dagberg B, de Haan LAM et al. The positive regulator CfaD overcomes the repression mediated by histone-like protein H-NS (H1) in the CFA/I fimbrial operon of *Escherichia coli*. EMBO J 1992; 11:2627-2632.

50. Harrison JA, Pickard D, Higgins CF et al. Role of *hns* in the virulence phenotype of pathogenic salmonellae. Mol Microbiol 1994; 13:133-140.

51. Spurio R, Durrenberger M, Falconi M et al. Lethal overproduction of the *Escherichia coli* nucleoid protein: ultramicroscopic and molecular autopsy. Mol Gen Genet 1992; 231:201-211.

52. Higgins CF, Hinton JCD, Hulton CSJ et al. Protein H1: a role for chromatin structure in the regulation of bacterial gene expression and virulence? Mol Microbiol 1990; 4:2007-2012.

53. Ueguchi C, Mizuno T. The *Escherichia coli* nucleoid protein H-NS functions directly as a transcriptional repressor. EMBO J 1993; 12:1039-1036.

54. Pruss GJ, Drlica K. DNA supercoiling and prokaryotic transcription. Cell 1989; 56:521-523.

55. Dorman CJ, Ni Bhrian N, Higgins CF. DNA supercoiling and environmental regulation of virulence gene expression in *Shigella flexneri*. Nature 1990; 344:789-792.

56. Hromockyj AE, Maurelli AT. Identification of an *Escherichia coli* gene homologous to *virR*, a regulator of *Shigella* virulence. J Bacteriol 1989; 171:2879-2881.

57. Rimsky S, Spassky A. Sequence determinants for H1 binding on *Escherichia coli lac* and *gal* promoters. Biochemistry 1990; 29:3765-3771.

58. Bracco L, Kotlarz D, Kolb A et al. Synthetic curved DNA sequences can act as transcriptional activators in *Escherichia coli*. EMBO J 1989; 8:4289-4296.

59. Tanaka K, Muramatsu S, Yamada H et al. Systematic characterization of curved DNA segments randomly cloned from *Escherichia coli* and their functional significance. Mol Gen Genet 1991; 226:367-376.

60. Lucht JM, Dersch P, Kempf B et al. Interactions of the nucleoid-associated DNA-binding protein H-NS with the regulatory region of the osmotically controlled *proU* operon of *Escherichia coli*. J Biol Chem 1994; 269:6578-6586.

61. Falconi M, Higgins P, Spurio R et al. Expression of the gene encoding the

major bacterial nucleoid protein H-NS is subject to transcriptional auto-repression. Mol Microbiol 1993; 10:273-282.

62. Dersch P, Schmidt K, Bremer E. Synthesis of the *Escherichia coli* K-12 nucleoid-associated DNA-binding protein H-NS is subjected to growth-phase control and autoregulation. Mol Microbiol 1993; 8:875-889.

63. Tobe T, Yoshikawa M, Mizuno T et al. Transcriptional control of the invasion regulatory gene *virB* of *Shigella flexneri*: activation by VirF and repression by H-NS. J Bacteriol 1993; 175:6142-6149.

64. La Teana A, Brandi A, Falconi M et al. Identification of a cold shock transcriptional enhancer of the *Escherichia coli* gene encoding nucleoid protein H-NS. Proc Natl Acad Sci USA 1991; 88: 10907-10911.

65. Rhode JR, Fox JM, Minnich SA. Thermoregulation in *Yersinia enterocolitica* is coincident with changes in DNA supercoiling. Mol Microbiol 1994; 12:187-199.

66. Falconi M, Gualtieri MT, La Teana A et al. Proteins from the prokaryotic nucleoid: primary and quaternary structure of the 15- kD *Escherichia coli* DNA binding protein H-NS. Mol Microbiol 1988; 2:323-329.

67. Kawula TH, Lelivelt KJ. Mutations in a gene encoding a new Hsp70 suppress rapid DNA inversion and *bgl* activation, but not *proU* derepression, in *hns*-1 mutant *Escherichia coli*. J Bacteriol 1994; 176:610-619.

68. Liu Q, Richardson CC. Gene 5.5 protein of bacteriophage T7 inhibits the nucleoid protein H-NS of *Escherichia coli*. Proc Natl Acad Sci USA 1993; 90:1761-1765.

69. Sakai T, Sasakawa C, Makino S et al. DNA sequence and product analysis of the *virF* locus responsible for congo red binding and cell invasion in *Shigella flexneri* 2a. Infect Immun 1986; 54:395-402.

70. Sakai T, Sasakawa C, Yoshikawa M. Expression of four virulence antigens of *Shigella flexneri* is positively regulated at the transcriptional level by the 30 kilodalton *virF* protein. Mol Microbiol 1988; 2:589-597.

71. Adler B, Sasakawa C, Tobe T et al. A dual transcriptional activation system for the 230 kb plasmid genes coding for virulence-associated antigens of *Shigella flexneri*. Mol Microbiol 1989; 3:627-635.

72. Hoe NP, Minion FC, Goguen JD. Temperature sensing in *Yersinia pestis*: regulation of *yopE* transcription by *lcrF*. J Bacteriol 1992; 174:4275-4286.

73. Gallegos MT, Michan C, Ramos JL. The XylS/AraC family of regulators. Nucleic Acids Res 1993; 21:807-810.

74. Higgins DG, Sharp PM. CLUSTAL: a package for performing multiple sequence alignment on a microcomputer. Gene 1988; 73:237-244.

75. Savelkoul PHM, Willshaw GA, McConnell MM et al. Expression of CFA/I fimbriae is positively regulated. Microbial Pathogen 1990; 8:91-99.

76. Caron J, Coffield L, Scott J. A plasmid-encoded regulatory gene, *rns*, required for expression of the CS1 and CS2 adhesins of enterotoxigenic *Escherichia coli*. Proc Natl Acad Sci USA 1989; 86:963-967.

77. Tobe T, Yoshikawa M, Mizuno T et al. Transcriptional control of the invasion regulatory gene *virB* of *Shigella flexneri*: activation by VirR and repression by H-NS. J Bacteriol 1993; 175:6142-6149.

78. Dagberg B, Uhlin BE. Regulation of virulence-associated plasmid genes in enteroinvasive *Escherichia coli*. J Bacteriol 1992; 174:7606-7612.

79. Buysse JM, Venkatesan M, Mills JA et al. Molecular characterization of a trans-acting, positive effector (*ipaR*) of invasion plasmid antigen synthesis in *Shigella flexneri* serotype 5. Microbial Pathogen 1990; 8:197-211.

80. Watanabe H, Arakawa E, Ito K et al. Genetic analysis of an invasion region by use of a Tn*3-lac* transposon and identification of second positive regulator gene, *invE*, for cell invasion of *Shigella sonnei*: significant homology of InvE with ParB of plasmid P1. J Bacteriol 1990; 172: 619-629.

81. Jost BH, Adler B. Site of transcriptional activation of *virB* on the large plasmid of *Shigella flexneri* 2a by VirF, a member of the AraC family of transcriptional activators. Microbial Pathogen 1993; 14:481-488.

82. Martin KA, Friedman SA, Austin SJ. Partition site of the P1 plasmid. Proc Natl Acad Sci USA 1987; 84:8544-8547.

83. Allaoui A, Sansonetti PJ, Parsot C. MxiD, an outer membrane protein necessary for the secretion of the *Shigella flexneri* Ipa invasins. Mol Microbiol 1993; 7:59-68.

CONTROL OF *LISTERIA MONOCYTOGENES* VIRULENCE BY THE TRANSCRIPTIONAL REGULATOR PRFA

Jürgen Kreft, Jutta Bohne, Roy Gross, Hubert Kestler,
Zeljka Sokolovic and Werner Goebel

1. INVASIVE *LISTERIA* SPECIES AND THEIR INTRACELLULAR LIFE STYLE

*L*isteria monocytogenes* is the causative agent of severe infections in humans and animals. These Gram-positive, rod-shaped and non-sporulating bacteria belong to the genus *Listeria* which comprises next to *L. monocytogenes*, the species *L. ivanovii* which is pathogenic for animals but not for humans and the four apathogenic species *L. seeligeri*, *L. innocua*, *L. grayi* and *L.welshimeri* which are frequently encountered in the natural environment.[1] Risk groups among humans that are primarily affected by *L. monocytogenes* are immunocompromised people and pregnant women. The major symptoms of listeriosis are septicemia, stillbirth and meningitis.[2] The bacteria seem to be taken up by contaminated food and the intestine appears to be the major route of entry in humans and animals.[3,4] After crossing the intestinal barrier these facultative intracellular bacteria are able to replicate in the cytoplasm of the infected epithelial cells after disrupting the endosomal membrane.[5] At least in established mammalian cell lines, *L. monocytogenes* spreads from cell to cell without leaving the intracellular environment. In these cell cultures it has been shown that the intra- and intercellular movement is brought about by the formation of polymerized actin filaments. These filaments are first formed on the surface of the listerial cell. They later rearrange to tail-like structures

Signal Transduction and Bacterial Virulence, edited by Rino Rappuoli,
Vincenzo Scarlato and Beatrice Aricò. © 1995 R.G. Landes Company.

at the old pole of the bacterial cell. Continuous polymerization is thought to be the propulsive force which moves the bacteria through the cytoplasm of the host cell whereby the actin tail grows proportionally (for a review see ref. 6). It is assumed that the subsequent uptake of the listeriae by macrophages occurs by a similar mechanism as the cell-to-cell spread between neighboring epithelial cells.[7] Thereby the macrophage engulfs the pseudopod-like structure which is formed when the listerial cell with the actin tail behind reaches the inside of the host cell's membrane. Monocytes may eventually transport the intracellular listeriae into the blood and further to the meningi. In the mouse model large amounts of viable listeriae are found within few hours post infection in the spleen and the liver.[8] Clearance of the intracellular bacteria occurs primarily by cytotoxic (CD8) and helper (CD4) T-cells (for a review see ref. 9 and 10).

2. VIRULENCE-ASSOCIATED GENES AND THEIR REGULATION BY PRFA

Several chromosomal genes, the products of which contribute to *L. monocytogenes* virulence have been identified.[11] Six of these genes, *prfA*, *plcA*, *hly*, *mpl*, *actA* and *plcB* form a cluster (Fig. 8.1) which seems to be localized on the chromosome

of all clinical isolates of *L. monocytogenes* analyzed so far between the *prs* gene coding for phosphoribosyl-pyrophosphate synthetase and the *ldh* gene encoding lactate dehydrogenase.[12,13] The genes *plcA* and *plcB* encode phospholipases of the C-type. Whereas PlcA possesses specificity for phosphatidyl inositol, PlcB preferentially cleaves phosphatidyl choline (lecithin).[14-16] *Hly* encodes listeriolysin,[17] *mpl* a zinc-dependent metalloprotease[18,19] and *actA* a protein which is required for the polymerization of F-actin in the cytoplasm of the infected host cell.[20,21] These gene products are essential for the intracellular life cycle of *L. monocytogenes*: the experimental data obtained mainly on mammalian cell cultures support the view, that listeriolysin and PlcA disrupt the phagosomal membrane,[5,22,23] ActA either directly or indirectly acts as nucleator for polymerization of F-actin[6] and PlcB is required for the lysis of the double membrane[16] which is formed when the listeriae spread between host cells. The function of Mpl is not yet clearly understood. It has been, however, demonstrated that Mpl is required for the conversion of the proform of PlcB into the mature enzymatically active form.[4] Mpl can also cleave ActA into several defined fragments (Sokolovic Z, unpublished data). However, it is unknown whether this reaction has any functional

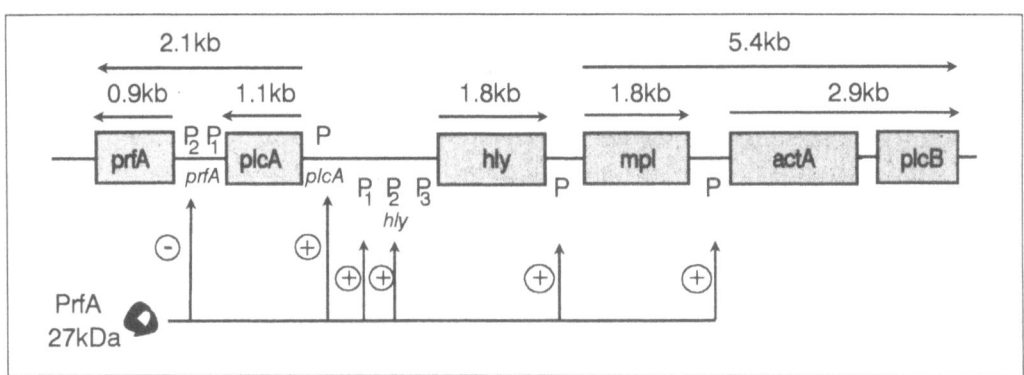

Fig. 8.1. Organization and transcriptional map of the PrfA-controlled gene cluster comprising the genes prfA, plcA, hly, mpl, actA and plcB. The arrows above the genes indicate the direction and length (in kilobases) of the observed transcripts. The arrows below point to promoters (P) which are positively (+) or negatively (−) regulated by PrfA. The prfA promoters P1/P2 and the hly promoters P1-3 are described in detail in Figure 8.4 and in the text.

significance for the intracellular life cycle of *L. monocytogenes.*

These virulence genes and probably also *inlA*, a gene encoding internalin, which is involved in the uptake of *L. monocytogenes* by epithelial cells,[24] are regulated by a transcriptional activator known as PrfA.[12,25] As already pointed out above, the *prfA* gene is part of the chromosomal virulence gene cluster and its transcription is auto-regulated by PrfA (see below).[26,27] The first indication for the regulation of these virulence genes by a common positive transcription factor came from the analysis of a class of spontaneously occurring non-hemolytic mutants. The mutation was identified as a deletion mapping about 1.6 kb upstream of the *hly* gene[28] and caused the complete loss of virulence when the mutant was tested in mice.[12] It was further shown that the mutation did not only abolish synthesis of listeriolysin but also that of several other proteins.[12,29,30] Cloning and sequencing of the chromosomal locus affected by the deletion finally led to the identification of the *prfA* (positive regulating factor A) gene.[26,27] Virulence of the *prfA* deletion mutant was partially restored by trans complementation with the *prfA* gene cloned on a multicopy plasmid.[27] Furthermore, transposon insertions and site-specific mutations in the *prfA* coding region or the *prfA* promoter led to the same phenotype as the spontaneous deletion mutation described above.[12,31] All of these mutations blocked completely the transcription of the entire gene cluster, i.e. *plcA*, *hly*, *mpl*, *actA* and *plcB*, indicating that PrfA is a positive transcription factor which is strictly required for the expression of these virulence genes.[12,27,31] In contrast to the strict dependence of the clustered virulence genes on PrfA, transcription of the internalin gene depends only partially on PrfA as will be discussed below.[25]

3. PRFA IS A MEMBER OF THE CRP/FNR FAMILY

The *prfA* gene of *L. ivanovii* has recently been cloned and sequenced.[32] Both the gene and its product share extensive homology (79% and 77% identity, respectively) with their *L. monocytogenes* counterparts. According to sequence similarities the PrfA proteins belong to the well known Crp-Fnr family of bacterial transcription factors (Fig. 8.2). With the exception of a regulatory protein of *Lactobacillus casei*, Crp (also called Cap) and its homologues have been found only in Gram-negative bacteria. Members of this family are involved in a variety of adaptive processes of these bacteria. For example, the *Escherichia coli* Crp protein primarily regulates the expression of genes involved in catabolic functions and is the central regulator of the catabolite repression. Fnr is a main regulator in the cellular response of *E. coli* to anaerobic growth conditions. Other prominent members of this fast growing family of regulatory proteins are the FixK protein of *Rhizobium meliloti* and the NtcA protein of *Synechococcus* spp. involved in the control of the nitrogen metabolism. So far, only two other proteins, HlyX of *Actinobacillus pleuropneumoniae* and Crp of *Xanthomonas campestris*, seem to be involved in the regulation of bacterial virulence.

One important question regarding the function of these proteins is, how their activity to function as transcription factors is regulated. Most is known about the function of the Crp protein in the catabolite repression (for a detailed review see ref. 33). The mediator for the catabolite repression is cAMP and in the presence of a preferred carbon source such as glucose, *E. coli* lowers the cellular cAMP concentration. The regulatory protein Crp has a high affinity for this nucleotide and becomes active only as a complex with cAMP. cAMP-Crp is able to activate the expression of genes, which are regulated negatively by glucose. However, Crp is involved not only in the activation of genes sensitive to catabolite repression, but expression of many different genes in *E. coli* is also regulated by this protein.

The three dimensional structure of the Crp protein is well known, because the

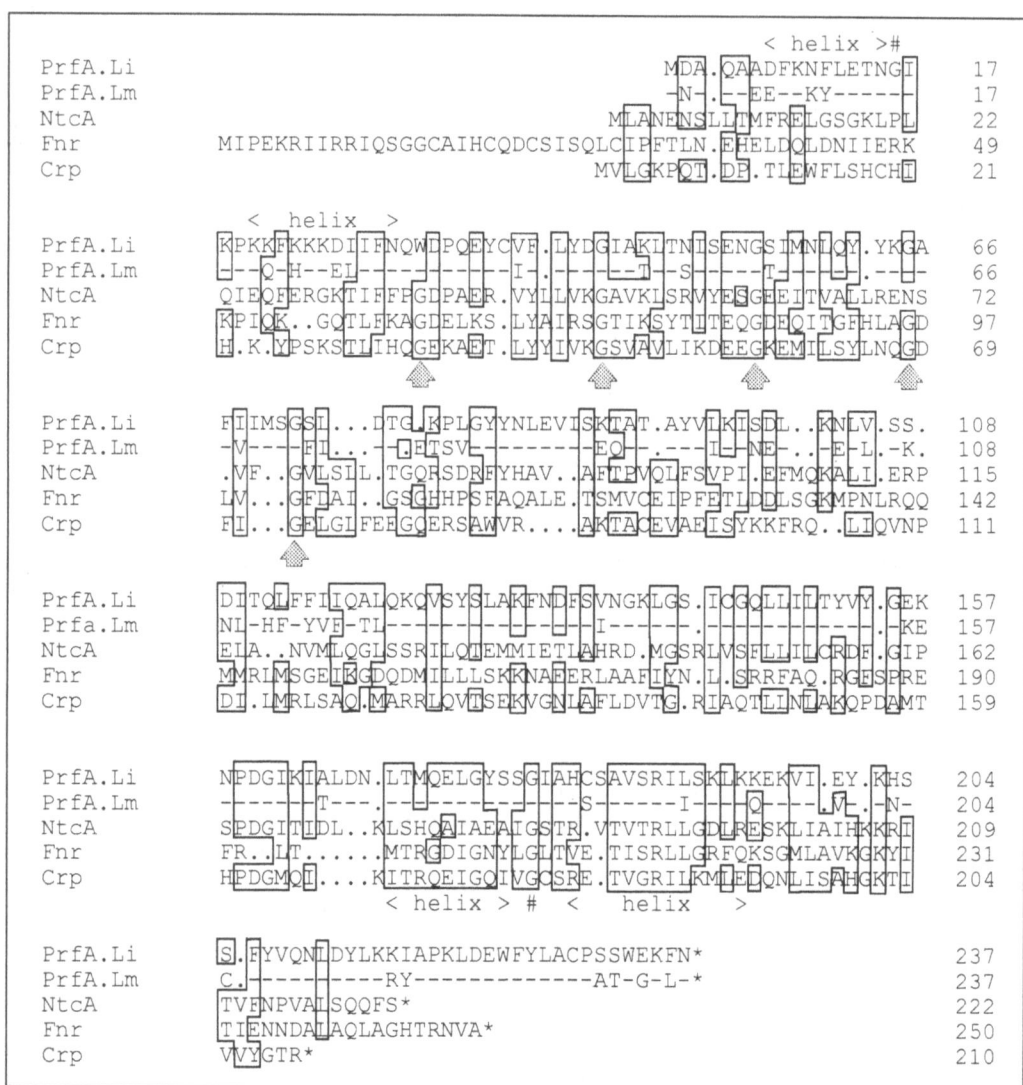

Fig. 8.2. Multiple amino acid sequence alignment of PrfA from L. ivanovii (PrfA.Li) and L. monocytogenes LO28 (PrfA.Lm), NtcA from Synechococcus, Fnr and Crp from E. coli (cited in ref. 32). The putative helix-turn-helix at the N-terminus of PrfA and the DNA-binding helix-turn-helix of Crp are indicated above or below the sequence, respectively. The glycine in the turn region is marked by #, highly conserved glycines in the N-terminal part are indicated by arrows. Amino acids which are similar in PrfA and in at least two of the other proteins are boxed. Groups of similar amino acids were as follows: (i) A, S, T, P, G; (ii)N, D, E, Q; (iii) H, R, K; (iv) M, L, I, V; (v) F, Y, W. Reprinted from Lampidis R et al, Mol Microbiol 1994; 13:141-151, with permission from Blackwell Science Ltd.

crystallization of the Crp-cAMP complex allowed the resolution of the structure at a resolution of 2.5 Å. Active Crp is a homodimer. Each Crp monomer is composed of two domains, an N-terminal nucleotide binding site and a C-terminal DNA binding domain containing a helix-turn-helix motif. Two cAMP molecules can be bound by a dimer and the nucleotide itself is directly involved in the establishment of the dimeric structure of Crp. A region forming β-roll structures, which are separated by several glycines, seems to be important for cAMP binding and dimerization. Although no crystallographic data are available for the Crp homologues, it is likely that the general organization of these proteins may be very similar to Crp, because many of the structural features mentioned above are strictly conserved among all of them. All of these proteins contain a C-terminal helix-turn-helix motif, which may be their DNA binding domain. As recently proposed by Fischer,[34] according to amino acid similarities in this motif the proteins can be classified into several subgroups, group I is represented by the Fnr and related proteins, group II by Crp's of various species, and group III by NtcA of *Synechococcus* spp, *Anabaena* etc. The β-roll structure separated by glycine residues is a typical structure of all proteins belonging to the Crp family. However, the Fnr protein and some related proteins such as HlyX of *Actinobacillus pleuropneumoniae* possess a domain at the N-terminus not present in the Crp protein. In this N-terminus three cysteines are present, and an additional cysteine is found in the central part of these proteins. In contrast to Crp, it seems that these proteins do not bind cAMP or other nucleotides as effector molecules, instead the cysteines may be involved in binding of metal ions such as iron, which could be the signal perceived by these proteins.

The PrfA proteins of *L. monocytogenes* and of the related species *L. ivanovii* can be included as new members in this Crp-Fnr family because of extensive sequence similarities and the conservation of the structural features described above.[32] Four of the glycines in the region of the β-roll structures are also found in PrfA at the corresponding sequence position and a high probability for short β-strands is also predicted. Similarly, the region of the helix-turn-helix motif shows very high sequence conservation. On the other hand, the central region of PrfA, which in the case of Crp is involved in nucleotide binding and in dimer formation, is less conserved on the amino acid level and consequently in the predicted secondary structure. It is difficult to classify the PrfA proteins into one of the three groups recently proposed

on the basis of sequence conservation. PrfA seems to be more closely related to the groups II and III (consisting of Crp and of NtcA-like proteins) than to the Fnr-like proteins (group I). Indeed, PrfA does not possess the putative N-terminal metal binding domain consisting of the cysteines typical for the Fnr-like proteins. Remarkably, PrfA has several unique features which distinguish this protein from all other members of this family. PrfA possesses a N-terminal helix-turn-helix domain in addition to the Crp-like helix-turn-helix motif close to the C-terminus. Furthermore, PrfA has a prolonged C-terminus as compared with the other proteins and this C-terminus includes a leucine-containing α-helix with structural properties resembling a leucine zipper. It is too early to speculate about the function of these PrfA specific domains, although mutations in the C-terminal region containing the putative leucine-zipper and in the Crp-like helix-turn-helix domain cause the inactivation of the protein (Lampidis R and Bubert A, unpublished data). No experimental data are available about possible effector molecules, which could interfere with PrfA activity. Although the nucleotide binding region present in Crp is less conserved in PrfA, the PrfA sequence does not allow final conclusions about nuleotide binding or other possible effectors interfering with its activity.

In several cases, it was shown that these regulatory proteins, mainly dependent on the position of the binding site in the promoter, can either activate or repress transcription of the respective genes. The mechanism of transcriptional activation mediated by Crp has been extensively investigated. In the absence of its effector cAMP, Crp exhibits a high but nonspecific DNA binding activity. The presence of cAMP favors the binding at specific sites with a defined palindromic consensus sequence. Frequently, DNA fragments carrying a specific Crp binding site display an anomalous mobility in agarose gels, suggesting that a bend is induced. The position of the Crp binding sites in

Crp-dependent promoters can vary from -40 to -200 with respect to the transcription start site. Generally, binding of Crp close to the RNA polymerase binding site indicates that Crp alone is sufficient for the activation of transcription. In the case of binding sites located further upstream, additional regulatory proteins are involved which act together with Crp in the activation of the respective promoters.

All known putative PrfA binding sites in the respective promoters of *L. monocytogenes* and *L. ivanovii* seem to be located in a region corresponding to the "-41 position" of the Crp-activated promoters. It is possible that their activation mechanism resembles that of the corresponding Crp-dependent promoters. Thus, there exist obvious parallels between PrfA and the Crp-family in the structure of the proteins and also in the position and sequence of DNA targets. It is certainly premature to speculate on such PrfA mediated mechanisms of transcriptional activation merely on the basis of these sequence homologies and differences. On the other hand, PrfA possesses several unique features which distinguishes it from the other Crp-like proteins, such as an additional helix-turn-helix (HTH) domain and a potential leucine zipper at its C-terminus. These features may have evolved for its specific function as the central virulence regulator in Listeria species. Further experiments are necessary to assign and verify functions for these features.

4. EXPRESSION SITES OF PrfA AND PrfA-REGULATED GENES

The expression of the PrfA regulator protein itself is controlled in *L. monocytogenes* in a complex way. In front of the *prfA* gene two promoters (P1 and P2) were identified both of which lead to a monocistronic *prfA* message of 0.9 kb and 0.8 kb, respectively.[35] An additional transcript (2.1 kb) containing the PrfA-coding sequence is synthesized from a promoter located in front of *plcA*. Transcription of this bicistronic mRNA is activated by PrfA[22,27] whereas transcription of both

monocistronic *prfA* mRNAs seems to be repressed by the PrfA protein.[36] The *prfA* promoters P1 and P2 appear to be functionally redundant at least in vivo. However, a double mutant lacking both promoters is 100-fold less virulent than the wild type and was delayed in the release of the listeriae from the phagosome. Synthesis of the bicistronic *prfA* transcript depends on the activity of at least one of these *prfA* promoters,[36] and is necessary for the full expression of the PrfA-dependent genes. It has been reported that synthesis of the monocistronic *prfA* transcripts occurs preferentially in the late logarithmic and early stationary growth phase, while the bicistronic transcript is mainly transcribed in the logarithmic growth phase.[27]

A palindromic sequence of 14 bp was identified in the upstream -35 region of the PrfA-dependent *hly* promoter. Nucleotide exchanges at specific positions of this sequence abolish entirely the transcription of the *hly* gene.[37] Similar sequences were also found in similar locations in all other promoters which are PrfA-controlled.[18,20,25,38] Single nucleotide exchanges are present in the *mpl* and *actA* palindromic sequences and the corresponding sequence in the *inlA* promoter is even more degenerated. Interestingly, in *L. ivanovii* the palindromic sequences preceding the respective genes in all cases are two base pairs longer than those found in *L. monocytogenes* (Fig. 8.3).[32] The relevance of this finding remains unclear at present. It was proposed that PrfA may directly bind to this sequence, sometimes referred to as "PrfA-box". Evidence that this may be indeed the case was provided by the work of Freitag et al[35] who carried out binding studies with a purified PrfA protein which contained, however, at its N-terminus 12 amino acids derived from the glutathione-S-transferase gene into which the *prfA* gene was inserted. Although the experiments suggested preferential binding to the palindromic sequence, large amounts of the fusion protein were required to achieve reasonable binding. The question remains

L.monocytogenes hly / plcA	CATTAACAAATGTTAAcG GTAATTGTTTACAATTgC	L.monocytogenes mpl	TTAACAAATGTaAA AATTGTTTACAtTT
L.ivanovii ilo / plcA	TCTTTAACAAATGTTAAAGA AGAAATTGTTTACAATTTCT	L.ivanovii mpl	TTTAACAAATGTcAAA AAATTGTTTACAgTTT
L. monocytogenes actA	TAACAAATGTTA ATTGTTTACAAT	L. monocytogenes inlA (P2)	TAACATAaGTTA ATTGTATtCAAT
L. ivanovii actA	TTAACAAATGTTAA AATTGTTTACAATT		

Fig. 8.3. Comparison of the "PrfA-boxes". The sequences for L. monocytogenes hly/plcA, mpl, actA, and inlA are from refs. 18, 20, 25, 38; the sequences for L. ivanovii ilo/plcA, mpl and actA are from ref. 32 and from our own unpublished results. Nonsymmetrical nucleotides are given in lower case. Reprinted from Lampidis R et al, Mol Microbiol 1994; 13:141-151, with permission from Blackwell Science Ltd.

therefore unanswered whether the inefficient binding is due to the additional N-terminal 12 amino acids in this PrfA fusion protein or whether efficient binding of PrfA requires additional factor(s).

Evidence for the assumption that PrfA might interact with another factor(s) comes from several experimental findings, most of which will be discussed in detail below. One observation supporting this notion is that the apathogenic species *L. seeligeri* synthesizes a cytolysin, termed seeligerolysin, which is very similar to listeriolysin from *L. monocytogenes*.[39] However, seeligerolysin is made in very low amounts when compared to listeriolysin. Recently it has been shown by hybridization that *L. seeligeri* contains a DNA sequence with homology to prfA from *L. monocytogenes*.[13] In addition, the seeligerolysin gene is preceded by a perfect consensus 14 bp "PrfA-box" (Kreft J et al, unpublished data). The low expression of seeligerolysin could be explained by a lack of functional PrfA. However, the introduction of multiple copies of the *L. monocytogenes prfA* gene into *L. seeligeri* did not result in an increased synthesis of seeligerolysin (Kreft J et al, unpublished data). A possible explanation for this observation is that *L. seeligeri* lacks a factor(s) which is present in the other two hemolytic Listeria species and which is necessary, in addition to PrfA, for efficient listeriolysin synthesis.

As described above, in *L. monocytogenes* the two prfA promoters, P1 and P2, are downregulated by higher levels of PrfA. It was shown that a 20 bp-deletion in the -35 region of P2 increases the *prfA* transcript from P1 about 20-fold.[36] Within the deleted part a 14 bp sequence can be identified which resembles a PrfA-binding site, except that one half of the palindrome is inverted to create a repeat on the opposite strand of the DNA helix. Freitag and Portnoy[36] proposed that binding of PrfA, which most presumably acts as a dimer, to this opposite strand repeat may distort PrfA in such a way that it no longer functions as an activator but rather as a repressor. This, however, seems unlikely since in this case either dimer formation of the PrfA protein or DNA-binding via the HTH in PrfA would exclude each other for sterical reasons. Our own analysis of the *prfA* promoter region identified an almost perfect "PrfA-box" which is only slightly shorter than the consensus sequence (Fig. 8.4). This box is centered at position -37 of the P2 promoter, which is half a helix turn before the position in PrfA activated promoters. We therefore favor the assumption that binding of PrfA to this site, downstream from P1 and overlapping with the -35 position of P2, exerts a negative effect on *prfA* transcription due to sterical hindrance of initiation of RNA polymerase binding to the P2 promoter. Such dual functions

```
     .         .         .         .         .         .
TAACTTATATTTTCCTATAAAAGGGTTAGTATATCTCCGAACCACATTCGGAAACATATA    60
>end (plcA)

                                       .         .         .
CTAACCCTTTTTCTATAGCATTTCAACTAGAAATAATAGCCTACAGATACTCCAGGAAAA   120

                           .       -----       ----         .
AAATTTTAATACGTAAGCAAATCTTTTGCGAATTTAAAAATTGTATAATAAAATCCTATA   180
                  -35 (P1)                     -10 (P1)
-------.         .         ----.         .TAACAATTGTTG       .
TGTAAAAAACATTATTTAGCGTGATTTTCTATCAACAGCTAACAATTGTTATTAGTGCCT   240
                                        -35 (P2)
         .     ----- -----       -----        .         .
AGCGCAATCATAAGGATAATTTTTCAAAAAGGCTATAAAAACGATTGGGGGATGAAAAATG   301
          -10 (P2)                            RBS      Met    prfA>
```

Fig. 8.4. *Nucleotide sequence of the intergenic region between prfA and plcA from* L. ivanovii. *Putative -10/-35 regions, the putative ribosome binding site (RBS) and the stem-loop (putative transcription terminator) distal from plcA are underlined. The putative promoters have been located in analogy to* L. monocytogenes.[36] *Regularly spaced (dA)4-5 and (dT)4-5 tracts are indicated by a broken line above the sequence, the palindrome downstream from the prfA promoter P1 and overlapping with P2 is in bold face. The corresponding palindrome from* L. monocytogenes *is shown above the sequence. Reprinted from Lampidis R et al, Mol Microbiol 1994; 13:141-151, with permission from Blackwell Science Ltd.*

of regulators are well known for other transcriptional activators[40-43] including Crp and Fnr, which, as pointed out above, share considerable sequence similarities with PrfA. This hypothesis is further supported by the recent finding that the shift of *L. monocytogenes* or *L. ivanovii* cultures from a rich growth medium (BHI) into Minimal Essential Medium (MEM) induces several PrfA-dependent proteins, at least one of which is negatively regulated by PrfA.[30,32]

The *prfA* promoter region differs from other PrfA-controlled virulence genes not only by the length and position of the "PrfA-box". In addition we could demonstrate experimentally that this DNA sequence, due to the presence of regularly spaced (dA)n- and (dT)n-tracts (Fig. 8.4), exhibits an intrinsic curvature.[32] These data are in good agreement with calculations of the intrinsic DNA curvature[44] for the relevant regions. Figure 8.5A shows that both the -10 box and the -35/"PrfA-box"-like sequence of P2 are located on the inner surface of a significantly bent DNA stretch. In contrast, the "PrfA-box" between *plcA* and *hly* as well as the respective -10 boxes are located on a DNA region

which shows no significant DNA bending (Fig. 8.5B). Similar to the case of Crp, such a DNA structure may have an effect on *prfA* transcription.

5. TRANSCRIPTIONAL ORGANIZATION

The transcriptional organization of the PrfA-controlled virulence genes has been studied in great detail. The *plcA* gene is cotranscribed with *prfA* in the above mentioned bicistronic mRNA and in an additional monocistronic mRNA.[12,27] The listeriolysin gene, *hly*, is transcribed from two PrfA-dependent promoters exclusively into a monocistronic mRNA.[12,27] A third *hly* promoter, also located in the intragenic region between *plcA* and *hly* downstream from the other two *hly* promoters, was recently identified. This promoter is independent of PrfA and results only in low level transcription of the *hly* gene.[45] The "PrfA-box" of the PrfA-dependent *hly* promoters seems to be identical to that of the *plcA* promoter but transcription of the two genes is divergent (Fig. 8.1). A polycistronic transcript is synthesized starting from a PrfA-regulated promoter in front of *mpl*. This transcript comprises the

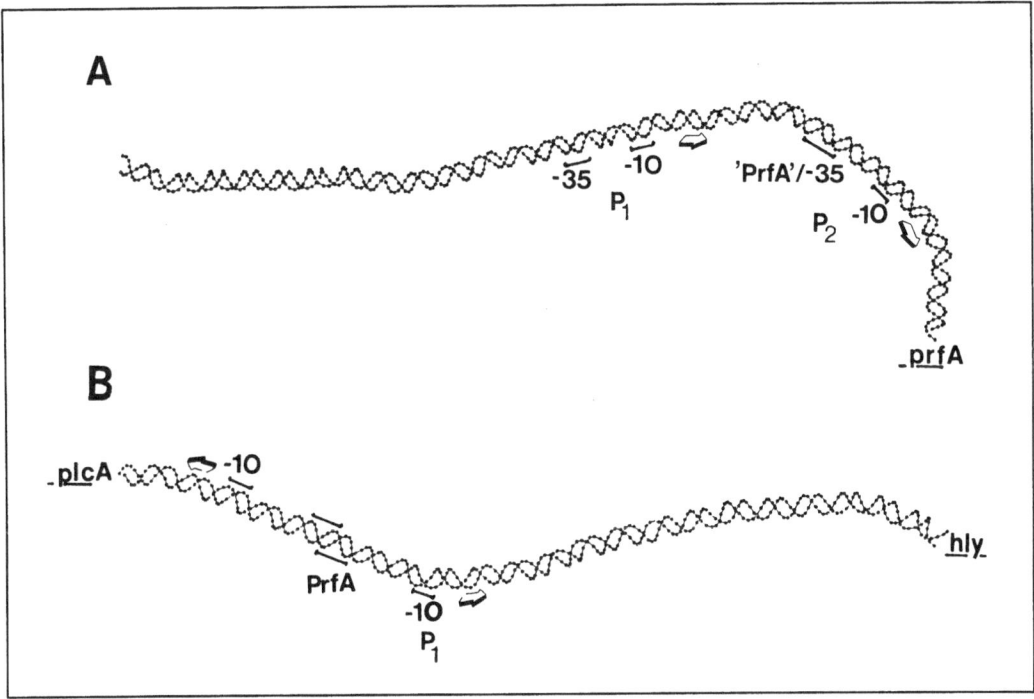

Fig. 8.5. Curvature of DNA, calculated according to Shpigelman et al,[44] in the prfA promoter region (A) and in the intergenic region between plcA and hly (B). 'PrfA'/-35 in (A) indicates the position of the "PrfA-box"-like sequence centered at pos. -37 of the prfA promoter P2 (see Fig. 8.4 and text); PrfA in (B) is the "PrfA-box" centered at pos. -41 of the divergently transcribed plcA and hly P1 promoters (see Figs. 8.1 and 8.3). Transcription start points are indicated by arrows.

coding sequences of *mpl*, *actA*, *plcB* and three additional ORFs with unknown function. Mpl may be also transcribed into a monocistronic mRNA.[12,21] Another PrfA-regulated promoter is located directly in front of the *actA* gene.[20] Transcription from this promoter leads to a bicistronic transcript which comprises the coding regions of *actA* and *plcB*.[46]

The transcription of *inlA*, also apparently dependent on PrfA, starts from several promoters located in front of this gene and involves a monocistronic mRNA and a bicistronic mRNA. The latter transcript comprises in addition to *inlA* the gene *inlB*, highly related to *inlA*, with which it forms an operon.[24,25] The determination of the life time of the PrfA-regulated transcripts showed that the mRNAs for *hly*, *actA* and *plcB* are unusually stable whereas the *prfA-*,*plcA-* and especially the *mpl* transcripts are rather unstable.[46]

6. PRFA-DEPENDENT GENES ARE DIFFERENTIALLY REGULATED

The requirement of different PrfA-dependent virulence factors in the various steps of the intracellular life cycle of *L. monocytogenes* suggests that the genes encoding these proteins are differentially regulated inside the host cell. It has been already shown that environmental parameters, such as temperature, stress conditions or certain metabolites affect expression of PrfA-dependent proteins differently.[30,46-49] Nothing is known until now about possible signals of the infected host which may influence the regulation of the PrfA-controlled genes. Recently, the preferential synthesis of some listerial gene products was demonstrated inside epithelial (Sokolovic Z, unpublished data) and phagocytic host cells.[46,50,51] It was also shown that ActA is synthesized in rather large amounts when

L. monocytogenes enters the cytoplasm of the infected host cell.[46]

Two models are currently discussed for the differential regulation of PrfA-controlled genes. Model I is based on the cellular amount of PrfA and the different binding affinity of PrfA to the various palindromic sequences in front of the PrfA-regulated genes. This model suggests that in an initial step transcription of *prfA* occurs via the two PrfA-independent *prfA* promoters, P1 and P2. This results in a limited amount of PrfA, sufficient to activate the high-affinity PrfA-dependent *hly* and *plcA* promoters thus allowing the synthesis of the *plcA-prfA* transcript which in turn leads to an enhanced PrfA synthesis. The higher cellular level of PrfA now activates the *mpl* and *actA* promoters which as described above seem to have a lower affinity to PrfA due to the nucleotide mismatches in their palindromic sequences.[35,36] The higher cellular level of PrfA also leads to the observed downregulation of the monocistronic *prfA* transcript, either by aberrant binding of PrfA to the distorted palindromic sequence in front of the *prfA* promoter P2 as proposed by Freitag and Portnoy[36] or by binding to the "PrfA-box" like sequence at the -35 position of P2, as suggested above.

Indeed, introduction of multiple copies of the palindromic sequences of the various PrfA-regulated promoters into *L. monocytogenes* reduces its ability to invade and replicate in mammalian cells in the expected way. The *hly-plcA* sequence has the strongest negative effect, followed by that of *actA* and of *mpl*. Even the introduction of a multicopy plasmid which carries a fortuitous "PrfA-box"-like palindromic sequence exerts a negative effect on the intracellular multiplication of *L. monocytogenes* (Kestler H, unpublished data). These data suggest that the palindromic sequences bind PrfA and thereby lower the amount of PrfA available for the PrfA-dependent transcription of the genes necessary for the intracellular life cycle. It is also in line with the differential regulation of PrfA-dependent genes by the cellular level of PrfA that in the presence of an excess of the palindromic *hly-plcA* sequence within the listerial cell transcription of *actA* and *mpl* is more reduced than that of *hly* and *plcA*. Interestingly, the growth rate of these *L. monocytogenes* strains remains unaffected when cultured in BHI media suggesting that PrfA does not control genes the products of which are essential for extracellular growth in rich media (Kestler H, unpublished data).

Model II postulates additional factors and/or modification(s) of the PrfA protein to explain the differential regulation of the PrfA-controlled genes. This model is based on several experimental observations where an increase in the expression of the *prfA* gene is not correlated with a concomitant increase in the expression of PrfA-dependent genes on the transcriptional and the translational levels:

a. It has been shown that the expression of the PrfA-dependent virulence genes in *L. monocytogenes* strain EGD is thermoregulated i.e. no transcripts of these genes are synthesized at 20°C whereas the monocistronic *prfA* mRNA is expressed.[48] Large amounts of this monocistronic *prfA* transcript are also observed in strain EGD when grown in rich media under anaerobic conditions, but there is again little concomitant transcription of the other PrfA-dependent genes occurring (Bohne J, unpublished data). Likewise, transcription of the PrfA-dependent genes is very low when a *prfA* mutant is complemented in trans by a multicopy plasmid carrying the *prfA* gene from strain EGD.[27] Again, there is a very large amount of monocistronic *prfA* mRNA synthesized in brain heart infusion, BHI (Bohne J, unpublished data). There is no reason to assume that the monocistronic *prfA* transcript is not efficiently translated since it contains the entire coding region for PrfA and the same ribosome binding site as the bicistronic *plcA-prfA* transcript.

b. Shift of *L. monocytogenes* NCTC 7973 from the rich BHI medium into Minimal Essential Medium (MEM) leads to the transcriptional induction of most PrfA-regulated virulence genes.[46] There are at least three different patterns of induction observed for the PrfA-dependent genes in *L. monocytogenes* NCTC 7973. The first one comprises the transcription of *plcA-prfA* and *hly* which is induced instantly. The second group of genes including *actA* and *plcB* is transcriptionally induced later. The third group is represented by the genes *mpl* and *inlA*. Transcription of these genes is hardly induced at all after shift into MEM or may be even downregulated as seems to be the case for *inlA* (Bohne J, unpublished data). There is always a corresponding increase of the gene products observed under the conditions which lead to an induction of the transcription of the virulence genes.

One has to take into account, however, that in all these studies the expression of *prfA* can be measured so far only on the transcriptional level but not on the translational level. A polyclonal antiserum raised against a PrfA-fusion protein, overexpressed in *E. coli*, could not detect PrfA protein in Listeria cell extracts (Kreft J and Gross R, unpublished data). This might indicate that the intracellular amount of PrfA is extremely low or/and that the protein is very unstable. Nevertheless, these and other unpublished data suggest that the differential regulation of the PrfA-dependent genes does not seem to depend on the cellular PrfA concentration only but may involve other factor(s) the synthesis of which is triggered by the various environmental parameters. These factor(s) may either modify the PrfA protein thereby altering the binding activity and/or specificity of PrfA or they may act in concert with PrfA to bring about differential expression of the various PrfA-dependent genes.

The final answer concerning the precise mechanism of the PrfA-mediated regulation of the virulence genes of *L. monocytogenes* and *L.ivanovii* awaits therefore further detailed molecular studies.

ACKNOWLEDGMENTS

The authors would like to thank A. Bubert, F. Engelbrecht, and R. Lampidis for the communication of unpublished results. The Deutsche Forschungsgemeinschaft (SFB165) and the Fonds der Chemischen Industrie are thanked for financial support.

REFERENCES

1. Jones D. Current classification of the genus Listeria. Abstr 11th Int Symp Probl Listeriosis 1992; Abstr 2:7-8.
2. Gellin BG, Broome CV. Listeriosis. J Amer Med Ass 1989; 261:1313-1320.
3. Farber JM, Peterkin PI. *Listeria monocytogenes*: a food-born pathogen. Microbiol Rev 1991; 55:476-511.
4. Poyart C, Abachin E, Razafimanantsoa I et al. The zinc metalloprotease of *Listeria monocytogenes* is required for maturation of phosphatidylcholine phospholipase C: direct evidence obtained by gene complementation. Infect Immun 1993; 61:1576-1580.
5. Gaillard J-L, Berche P, Mounier J et al. In vitro model of penetration and intracellular growth of *Listeria monocytogenes* in the human enterocyte-like cell line Caco-2. Infect Immun 1987; 55:2822-2829.
6. Cossart P, Kocks C. The actin-based motility of the facultative intracellular pathogen *Listeria monocytogenes*. Mol Microbiol 1994; 13:395-402.
7. Tilney LG, Portnoy DA. Actin filaments and the growth, movement and spread of the intracellular bacterial parasite *Listeria monocytogenes*. J Cell Biol 1989; 109:1597-1608.
8. Mackaness GB. Cellular resistance to infection. J Exp Med 1962; 116:381-406.
9. Cossart P, Mengaud J. *Listeria monocytogenes* a model system for the molecular study of intracellular parasitism. Mol Biol Med 1989; 6:463-474.
10. Kaufmann SHE. Immunity to intracellular bacteria. Annu Rev Immunol 1993; 11:129-163.
11. Portnoy DA, Chakraborty T, Goebel W et al. Molecular determinants of *Listeria*

monocytogenes pathogenesis. Infect Immun 1992; 60:1263-1267.

12. Chakraborty T, Leimeister-Wächter M, Domann E et al. Coordinate regulation of virulence genes in *Listeria monocytogenes* requires the product of the *prfA* gene. J Bacteriol 1992; 174:568-574

13. Gouin E, Mengaud M, Cossart P. The virulence gene cluster of *Listeria monocytogenes* is also present in *Listeria ivanovii*, an animal pathogen, and *Listeria seeligeri*, a non-pathogenic species. Infect Immun 1994; 8:3550-3553.

14. Leimeister-Wächter M, Domann E, Chakraborty T. Detection of a gene encoding a phosphatidylinositol-specific phospholipase C that is coordinately expressed with listeriolysin in *Listeria monocytogenes*. Mol Microbiol 1991; 5:361-366.

15. Mengaud J, Braun-Breton C, Cossart P. Identification of a phosphatidylinositol-specific phospholipase C activity in *Listeria monocytogenes*: a novel type of virulence factor. Mol Microbiol 1991; 5:367-372.

16. Vazquez-Boland J-A, Kocks C, Dramsi S et al. Nucleotide sequence of the lecithinase-operon of *Listeria monocytogenes* and possible role of lecithinase in cell-to-cell spread. Infect Immun 1992; 60:219-230.

17. Mengaud J, Chenevert J, Geoffroy C et al. Identification of the structural gene encoding the SH-activated hemolysin of *Listeria monocytogenes*: Listeriolysin O is homologous to Streptolysin O and Pneumolysin. Infect Immun 1987; 55:3225-3227.

18. Domann E, Leimeister-Wächter M, Goebel W et al. Molecular cloning, sequencing, and identification of a metalloprotease gene from *Listeria monocytogenes* that is species specific and physically linked to the listeriolysin gene. Infect Immun 1991; 59:65-72.

19. Mengaud J, Geoffroy C, Cossart P. Identification of a new operon involved in *Listeria monocytogenes* virulence: its first gene encodes a protein homologous to bacterial metalloproteases. Infect Immun 1991; 59:1043-1049.

20. Domann E, Wehland J, Rohde M et al. A novel bacterial virulence gene in *Listeria monocytogenes* required for host cell microfilament interaction with homology to the proline rich region of vinculin. EMBO J 1992; 11:1981-1990.

21. Kocks C, Gouin E, Tabouret M et al. *Listeria monocytogenes*-induced actin assembly requires the *actA* gene product, a surface protein, Cell 1992; 68:521-531.

22. Camilli A, Tilney LG, Portnoy DA. The dual roles of *plcA* in *Listeria monocytogenes* pathogenesis. Mol Microbiol 1993; 8:143-157.

23. Portnoy DA, Jacks PS, Hinrichs DJ. Role of hemolysin for the intracellular growth of *Listeria monocytogenes*. J Exp Med 1988; 167:1459-1471.

24. Gaillard J-L, Berche P, Frehel C et al. Entry of *Listeria monocytogenes* into cells is mediated by internalin, a repeat protein reminiscent of surface antigens from Gram-positive cocci. Cell 1991; 65:1127-1141.

25. Dramsi S, Kocks C, Forestier C et al. Internalin-mediated invasion of epithelial cells by *Listeria monocytogenes* is regulated by the bacterial growth state, temperature and the pleiotropic activator PrfA. Mol Microbiol 1993; 9: 931-941.

26. Leimeister-Wächter M, Haffner C, Domann E et al. Identification of a gene that positively regulates expression of listeriolysin, the major virulence factor of *Listeria monocytogenes*. Proc Natl Acad Sci USA 1991; 87:8336-8340.

27. Mengaud J, Dramsi S, Gouin E et al. Pleiotropic control of *Listeria monocytogenes* virulence factors by a gene that is autoregulated. Mol Microbiol 1991; 5:2273-2283.

28. Leimeister-Wächter M, Goebel W, Chakraborty T. Mutations affecting hemolysin production in *Listeria monocytogenes* located outside the listeriolysin gene. FEMS Microbiol Lett 1989; 65: 23-30.

29. Sokolovic Z, Fuchs A, Goebel W. Synthesis of species-specific stress proteins by virulent strains of *Listeria monocytogenes*. Infect Immun 1990; 58: 3582-3587.

30. Sokolovic Z, Riedel J, Wuenscher M et al. Surface-asociated, Prfa-regulated proteins of *Listeria monocytogenes* synthesized under stress conditions. Mol Microbiol 1993; 8:219-227.

31. Barry RA, Bouwer HGA, Portnoy DA et al. Pathogenicity and immunogenicity of *Listeria monocytogenes* small-plaque mutants defective for intracellular growth and cell-to-cell spread. Infect Immun 1992; 60:1625-1632.

32. Lampidis R, Gross R, Sokolovic Z et al. The virulence regulator protein of *Listeria ivanovii* is highly homologous to PrfA from *Listeria monocytogenes* and both belong to the Crp-Fnr family of transcription regulators. Mol Microbiol 1994; 13:141-151.

33. Kolb A, Busby S, Buc H et al. Transcriptional regulation by cAMP and its receptor protein. Annu Rev Biochem 1993: 62:749-795.

34. Fischer HM. Genetic regulation of nitrogen fixation in Rhizobia. Microbiol Rev 1994; 58:352-386.

35. Freitag NE, Rong L, Portnoy DA. Regulation of the *prfA* transcriptional activator of *Listeria monocytogenes*: multiple promoter elements contribute to intracellular growth and cell-to-cell spread. Infect Immun 1993; 61:2537-2544.

36. Freitag NE, Portnoy DA. Dual promoters of the *Listeria monocytogenes prfA* transcriptional activator appear essential in vitro but are redundant in vivo. Mol Microbiol 1994; 12:845-853.

37. Freitag NE, Rong L, Portnoy DA. Transcriptional activation of the *Listeria monocytogenes* hemolysin gene in *Bacillus subtilis*. J Bacteriol 1992; 174:1293-1298.

38. Mengaud J, Vicente MF, Cossart P. Transcriptional mapping and nucleotide sequence of the *Listeria monocytogenes hlyA* region reveal structural features that may be involved in regulation. Infect Immun 1989; 57:3695-3701.

39. Haas A, Dumbsky M, Kreft J. Listeriolysin genes: Complete sequence of *ilo* from *Listeria ivanovii* and *lso* from *Listeria seeligeri*. Biochem Biophys Acta 1992; 1130:81-84.

40. Eiglmeier K, Honore N, Luchi S et al. Molecular genetic analysis of Fnr-dependent promoters. Mol Microbiol 1989; 3:869-878.

41. Forsman K, Goransson M, Uhlin BE. Autoregulation and multiple DNA interactions by a transcriptional regulatory protein in *Escherichia coli* pili biogenesis. EMBO J 1989; 8:1271-1277.

42. Urbanowski ML, Stauffer GV. Genetic and biochemical analysis of the MetR activator-binding site in the *metE metR* control region of *Salmonella typhimurium*. J Bacteriol 1989; 171:5620-5629.

43. Wek RC, Hatfield TW. Transcriptional activation at adjacent operators in the divergent overlapping *ilvY* and *ilvC* promoters of *Escherichia coli*. J Mol Biol 1988; 203:643-663.

44. Shpigelman ES, Trifonov EN, Bolshoy A. CURVATURE: software for the analysis of curved DNA. CABIOS 1993; 9:435-440.

45. Domann E, Wehland J, Niebuhr K et al. Detection of a *prfA*-independent promoter responsible for listeriolysin gene expression in mutant *Listeria monocytogenes* strains lacking the PrfA regulator. Infect Immun 1993; 61:3073-3075.

46. Bohne J, Sokolovic Z, Goebel W. Transcriptional regulation of *prfA* and PrfA-regulated virulence genes in *Listeria monocytogenes*. Mol Microbiol 1994; 11:1141-1150.

47. Datta AG, Kothary MH. Effects of Glucose, growth temperature, and pH on Listeriolysin O production in *Listeria monocytogenes*. Appl Environ Microbiol 1993; 59:3495-3497.

48. Leimeister-Wächter M, Domann E, Chakraborty T. The expression of virulence genes in *Listeria monocytogenes* is thermoregulated. J Bacteriol 1992; 174: 947-952.

49. Park SF, Kroll RG. Expression of listeriolysin and phosphatidylinositol-specific phospholipase C is repressed by the plant-derived molecule cellobiose in *Listeria monocytogenes*. Mol Microbiol 1993; 8:653-661.

50. Brundage RA, Smith GA, Camilli A et

al. Expression and phosphorylation of the *Listeria monocytogenes* ActA protein in mammalian cells. Proc Natl Acad Sci USA 1993; 90:11890-11894.

51. Klarsfeld AD, Goossens L, Cossart P. Five *Listeria monocytogenes* genes preferentially expressed in infected mammalian cells: *plcA, purH, purD, pyrE* and an arginine ABC transporter gene, *arpJ*. Mol Microbiol 1994; 13:585-597.

CHAPTER 9

SIGNAL TRANSDUCTION IN STAPHYLOCOCCI AND OTHER GRAM-POSITIVE COCCI

Richard P. Novick

1. INTRODUCTION

There is every reason to think that signal transduction is as frequent and important a regulatory mechanism in Gram-positive cocci (GPC) as it is in other more intensively analyzed bacteria such as *Escherichia coli* and *Bacillus subtilis*.

To date, only a few signal transduction systems have been identified in GPC and in only one of these has the entire pathway been clearly delineated. These systems are *agr*, a global regulator of virulence in staphylococci, *bla*, the induction system for staphylococcal β-lactamase,[1-3] *VanA*, the induction system for vancomycin (Vn) resistance in enterococci,[4] and the biosynthetic pathways for the lanthionine-containing bacteriocins such as nisin, subtilin and epidermin, produced by *Lactobacillus lactis*, *B. subtilis* and *Staphylococcus epidermidis*, respectively.[1] In addition, Bayles et al[5] have identified a number of putative signal transduction systems in staphylococci by probing for conserved sequences corresponding to the well-known motifs characteristic of the classical signal transduction proteins. One of the putative histidine phosphokinase (HPK) genes thus identified has been insertionally inactivated, resulting in gross clumping of the bacteria in culture and a tight compact colonial morphology. Other genes identified by this screen encode products resembling transport proteins, suggesting that signal transduction may be involved in certain transport pathways (Bayles K, personal communication). Several streptococcal virulence factor genes are transcriptionally upregulated by the *mry/virR* gene, which resembles the response regulators; however, no HPK component has been identified to date.

Signal Transduction and Bacterial Virulence, edited by Rino Rappuoli, Vincenzo Scarlato and Beatrice Aricò. © 1995 R.G. Landes Company.

Finally, the plasmid-determined conjugation systems of the enterococci respond to small peptides (pheromones) produced by potential recipient organisms. These pheromones are taken up by the peptide permease system and interact with intracellular proteins to activate the conjugation system. Thus, these systems might be more accurately considered as involving signal transport rather than signal transduction. Nevertheless, it seems appropriate to include them since they represent an important response to an external signal and, moreover, may play a role in virulence.

It is axiomatic that in bacteria signal transduction systems regulate genes that are involved in accessory functions, i.e., are not required for the basic processes of cell growth and multiplication. Nevertheless, it is remarkable that most of the known signal transduction systems in Gram-positive cocci are carried by mobile genetic elements such as plasmids and transposons (see Table 9.1), which means their presence is variable among strains of a particular species and implies that they have been imported from other species where they may have originated as housekeeping genes. The only signal transduction pathways that are not carried by variable genetic elements are the virulence regulators, *agr* in *S. aureus* and *mry/virR* in *S. pyogenes*. Nevertheless, many of the genes that are controlled by these regulators, at least in *S. aureus*, are themselves carried by variable genetic elements and it is remarkable that these genes, imported

Table 9.1. Signal transduction in Gram-positive cocci

Species[1]	System	Location[2]	Regulatory genes		Signaling ligand
			Sensor	Regulator	
Sa	β-lactamase	Pl, Tn	blaR1	blaR2 (?)	β-lactams
Sa	agr	Chr	agrA	agrC	pheromone
Sc	epidermin	Pl		epiQ	unknown
Ef	VanA	Pl, CT	vanR	vanS	vancomycin
Ef	conjugation	Pl, CT		traE1	pheromone
Ll	nisin	Pl, CT	nisR	nisK	nisin
Spy	virulence	Chr		mry/vir	CO_2 (?)

[1] Sa = *Staphylococcus aureus*; Se = *S. epidermidis*; Ef = *Enterococcus faecalis*; Ll = *Lactobacillis lactis*;
Spy = *Streptococcus pyogenes*
[2] Pl = plasmid; Tn = transposon; CT = conjugative transposon

Fig. 9.1 (opposite). Signal transduction circuits in Gram-positive cocci. Genes are represented by directional boxes with letters, major transcripts by zigzag lines and major promoters by "P". Gray arrows represent synthesis, activation, and function of various gene products. Trans-membrane receptors are depicted as filled-in shapes with a white internal arrow representing transduction of the signal.

A. The VanA system. Ligand, vancomycin, "Vn", shown bound to receptor, VanS, which transfers phosphate group to response regulator, VanR, which then acts on the VanHAXYZ promoter.

B. The Bla system. Ligand, β-lactam, is shown bound to receptor, BlaR1, which is thus caused (presumably) to activate the hypothetical response regulator, BlaR2, whose activated form, BlaR2*, would block the action of the repressor, BlaI, on the two divergent promoters, P-blaR and P-blaZ, thus inducing both transcripts.

C. The agr system. AgrD and B either encode or regulate the synthesis of a peptide ("activator"), which is hypothesized to bind to the putative agr receptor, AgrC, which would then activate the response regulator, AgrA, by the standard phosphotransfer mechanism. AgrA~P would then activate transcription from the divergent promoters, P2 and P3. An accessory factor, Sar, is also required at this stage. The P3 transcript, RNAIII, then activates transcription of the class I exoprotein genes and represses transcription of the class II, probably requiring accessory factors for both types of action. Translation of the RNAIII-encoded hld (δ-hemolysin) gene is presumed to unfold RNAIII, exposing sequences that pair with the nascent hla leader, which would otherwise fold into an untranslatable configuration.

Receive a FREE BOOK of your choice

Please help us out—Just answer the questions below, then select the book of your choice from the list on the back and return this card.

R.G. Landes Company publishes five book series: *Medical Intelligence Unit, Molecular Biology Intelligence Unit, Neuroscience Intelligence Unit, Tissue Engineering Intelligence Unit* and *Biotechnology Intelligence Unit*. We also publish comprehensive, shorter than book-length reports on well-circumscribed topics in molecular biology and medicine. The authors of our books and reports are acknowledged leaders in their fields and the topics are unique. Almost without exception, there are no other comprehensive publications on these topics.

Our goal is to publish material in important and rapidly changing areas of bioscience for sophisticated scientists. To achieve this goal, we have accelerated our publishing program to conform to the fast pace in which information grows in bioscience. Most of our books and reports are published within 90 to 120 days of receipt of the manuscript.

Please circle your response to the questions below.

1. We would like to sell our *books* to scientists and students at a deep discount. But we can only do this as part of a prepaid subscription program. The retail price range for our books is $59-$99. Would you pay $196 to select four *books* per year from any of our Intelligence Units–$49 per book–as part of a prepaid program?

                Yes   **No**

2. We would like to sell our *reports* to scientists and students at a deep discount. But we can only do this as part of a prepaid subscription program. The retail price range for our reports is $39-$59. Would you pay $145 to select five *reports* per year–$29 per report–as part of a prepaid program?

                Yes   **No**

3. Would you pay $39–the retail price range of our books is $59-$99–to receive any single book in our Intelligence Units if it is spiral bound, but in every other way identical to the more expensive hardcover version?

                Yes   **No**

To receive your free book, please fill out the shipping information below, select your free book choice from the list on the back of this survey and mail this card to:

  R.G. Landes Company, 909 S. Pine Street, Georgetown, Texas 78626 U.S.A.

Your Name _____

Address _____

City _____ State/Province: _____

Country: _____ Postal Code: _____

My computer type is Macintosh _____ ; IBM-compatible _____ ; Other _____

Do you own _____ or plan to purchase _____ a CD-ROM drive?

AVAILABLE FREE TITLES

Please check three titles in order of preference.
Your request will be filled based on availability. Thank you.

☐ Water Channels
Alan Verkman,
University of California-San Francisco

☐ The Na,K-ATPase:
Structure-Function Relationship
J.-D. Horisberger, University of Lausanne

☐ Intrathymic Development of T Cells
J. Nikolic-Zugic,
Memorial Sloan-Kettering Cancer Center

☐ Cyclic GMP
Thomas Lincoln, University of Alabama

☐ Primordial VRM System and the Evolution
of Vertebrate Immunity
John Stewart, Institut Pasteur-Paris

☐ Thyroid Hormone Regulation
of Gene Expression
Graham R. Williams, University of Birmingham

☐ Mechanisms of Immunological Self Tolerance
Guido Kroemer, CNRS Génétique Moléculaire et
Biologie du Développement-Villejuif

☐ The Costimulatory Pathway
for T Cell Responses
Yang Liu, New York University

☐ Molecular Genetics of Drosophila Oogenesis
Paul F. Lasko, McGill University

☐ Mechanism of Steroid Hormone Regulation
of Gene Transcription
M.-J. Tsai & Bert W. O'Malley, Baylor University

☐ Liver Gene Expression
François Tronche & Moshe Yaniv,
Institut Pasteur-Paris

☐ RNA Polymerase III Transcription
R.J. White, University of Cambridge

☐ src Family of Tyrosine Kinases in Leukocytes
Tomas Mustelin, La Jolla Institute

☐ MHC Antigens and NK Cells
Rafael Solana & Jose Peña,
University of Córdoba

☐ Kinetic Modeling of Gene Expression
James L. Hargrove, University of Georgia

☐ PCR and the Analysis of the T Cell Receptor
Repertoire
Jorge Oksenberg, Michael Panzara & Lawrence
Steinman, Stanford University

☐ Myointimal Hyperplasia
Philip Dobrin, Loyola University

☐ Transgenic Mice as an In Vivo Model
of Self-Reactivity
David Ferrick & Lisa DiMolfetto-Landon,
University of California-Davis and Pamela Ohashi,
Ontario Cancer Institute

☐ Cytogenetics of Bone and Soft Tissue Tumors
Avery A. Sandberg, Genetrix & Julia A. Bridge ,
University of Nebraska

☐ The Th1-Th2 Paradigm and Transplantation
Robin Lowry, Emory University

☐ Phagocyte Production and Function Following
Thermal Injury
Verlyn Peterson & Daniel R. Ambruso,
University of Colorado

☐ Human T Lymphocyte Activation Deficiencies
José Regueiro, Carlos Rodríguez-Gallego
and Antonio Arnaiz-Villena,
Hospital 12 de Octubre-Madrid

☐ Monoclonal Antibody in Detection and
Treatment of Colon Cancer
Edward W. Martin, Jr., Ohio State University

☐ Enteric Physiology of the Transplanted Intestine
Michael Sarr & Nadey S. Hakim, Mayo Clinic

☐ Artificial Chordae in Mitral Valve Surgery
Claudio Zussa, S. Maria dei Battuti Hospital-Treviso

☐ Injury and Tumor Implantation
Satya Murthy & Edward Scanlon,
Northwestern University

☐ Support of the Acutely Failing Liver
A.A. Demetriou, Cedars-Sinai

☐ Reactive Metabolites of Oxygen and Nitrogen
in Biology and Medicine
Matthew Grisham, Louisiana State-Shreveport

☐ Biology of Lung Cancer
Adi Gazdar & Paul Carbone,
Southwestern Medical Center

☐ Quantitative Measurement
of Venous Incompetence
Paul S. van Bemmelen, Southern Illinois University
and John J. Bergan, Scripps Memorial Hospital

☐ Adhesion Molecules in Organ Transplants
Gustav Steinhoff, University of Kiel

☐ Purging in Bone Marrow Transplantation
Subhash C. Gulati,
Memorial Sloan-Kettering Cancer Center

☐ Trauma 2000: Strategies for the New Millennium
David J. Dries & Richard L. Gamelli,
Loyola University

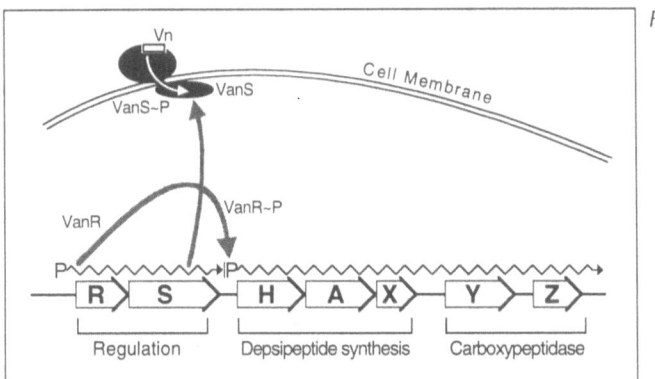

Fig. 9.1A

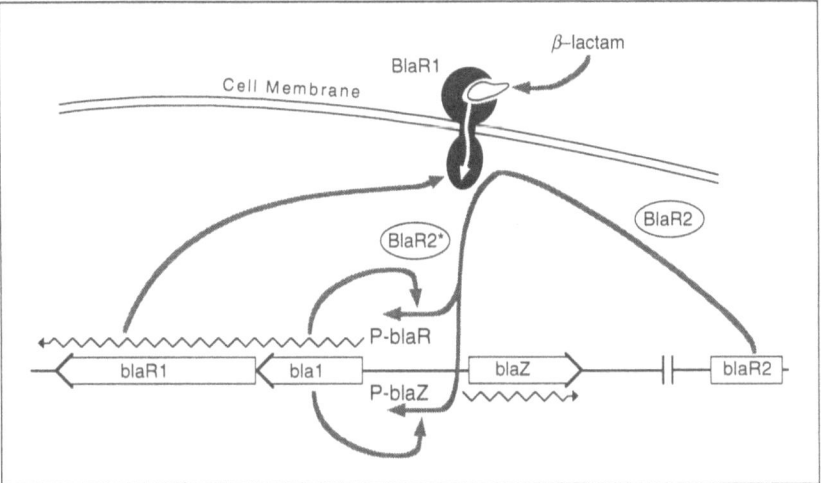

Fig. 9.1B

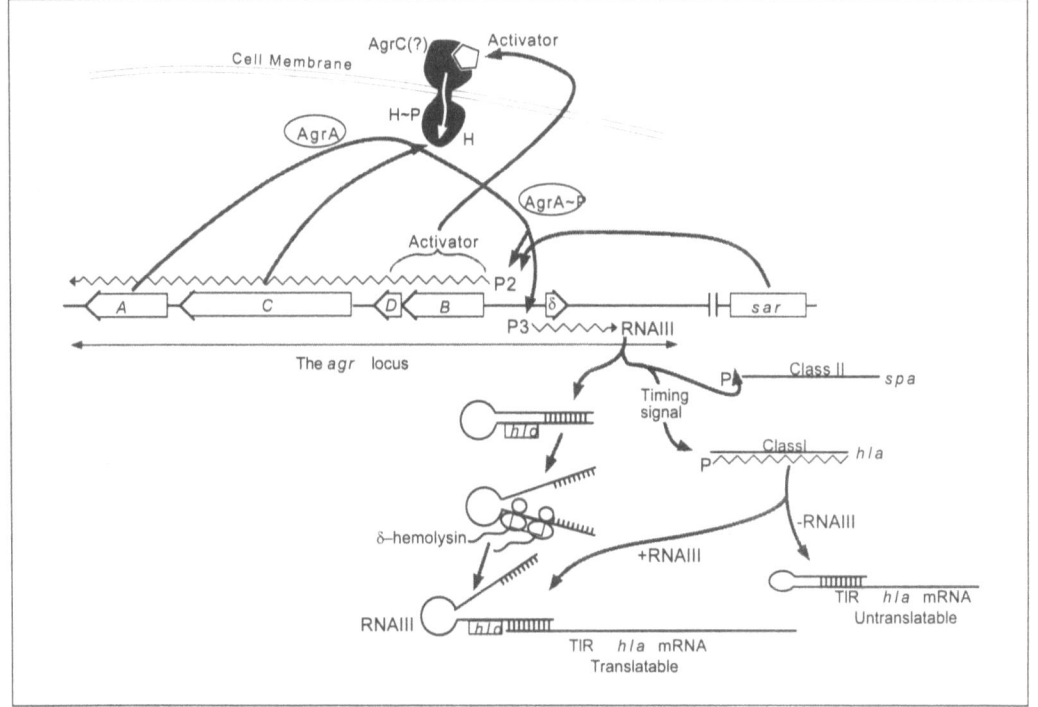

Fig. 9.1C

from other species, have come under the regulation of the *agr* system. The evolutionary process underlying this remarkable feature of staphylococcal virulence genes continues to be a great unsolved mystery.

2. THE VANA SYSTEM

Of the several systems, the plasmid-linked VanA (Fig. 9.1A) conforms most closely to the classical two-component systems that have been thoroughly characterized in Gram-negative bacteria. Here, the signal receptor is VanS which is assumed to respond to the extracellular ligand, vancomycin, by activating histidine phosphokinase activity and then transferring the phosphate group to VanR, which would activate the common promoter for *vanH*, *A*, *X*, *Y* and *Z*. Genetic data have clearly defined the activation sequence; the phosphotransfer mechanism remains hypothetical and is based on comparison of VanRS sequences with those of the known signal transduction components.

The organization of the regulon is unusual in that the regulatory genes, *vanR* and *vanS* are transcribed in the same direction as the target operon, VanHAXYZ. In general, the regulatory operon of any signal transduction system must be continually transcribed at a low basal rate to ensure the presence of sufficient levels of the regulatory proteins to enable the initial response to the external ligand. Probably for this reason, most systems in which the regulatory and target operons are linked have the two operons transcribed divergently so that there can be no transcriptional read-through. With the VanA system, therefore, there would have to be a strong transcriptional terminator for *vanRS* and, indeed, such a terminator is present. [4] In some systems, such as *agr* and the nisin pathway (see below) the signal transduction pathway activates its own transcription as well as that of its target genes, and is thus autocatalytic. In the VanA system, the signal transduction pathway apparently regulates only its target genes, namely the *vanHAXYZ* operon, whose gene products combine to make the

cell wall of the organism indifferent to the presence of Vn. The VanA system consists of several genes acting to protect cell wall biosynthesis from inhibition by Vn. The net effect of the system is to substitute a D-alanine-D-lactate depsipeptide for the standard D-ala-D-ala as terminus of the peptidoglycan precursor to which vancomycin binds and thus inhibits polymerization of the mucopeptide subunits.[6-8] In *Enterococcus faecalis*, the depsipeptide is not found in the mature peptidoglycan and is removed by the *vanY*-encoded D, D-carboxypeptidase (Courvalin P, personal communication), which also removes D-ala from any peptidoglycan synthesized by the endogenous chromosomal synthase system.

VanS is a trans-membrane protein that contains the classical histidine phosphokinase motifs of other bacterial signal transduction receptors.[9] It is assumed that the N-terminal region contains a Vn binding site; however this site has not been identified as yet. VanR, the response regulator, would be phosphorylated by VanS and then serve as a transcriptional activator of the *vanHAXYZ* promoter. The precise binding site is unknown; however, a gene fusion containing only 90 nt of the intergenic region is activated by vancomycin through the VanS-VanR pathway.[4]

As is the case with most antibiotic resistance in clinically relevant bacteria, the VanA regulon in the enterococci has been acquired from some other species where it presumably evolved as part of the normal genetic complement, later becoming associated with the transmissible elements (a conjugative transposon usually attached to a plasmid[10,11] that brought it to the enterococci). In the case of vancomycin, a likely source is one or another of the naturally Vn^r Gram-positive species such as lactobacilli, leuconostoc, etc. In these species, cell wall peptidoglycan is naturally terminated by a depsipeptide (D-ala-D-lactate), accounting for their intrinsic vancomycin resistance.[12] If this idea is correct, the fascinating question remains of how the peptidoglycan synthetic genes of such a species became associated with a signal transduction pathway responsive to vancomycin, on their way to residence in the enterococci. Since no homology can be detected by Southern blotting between genes of the VanA system and chromosomal DNA of these naturally resistant organisms, the putative gene transfer must have occurred a very long time ago, if at all.

3. β-LACTAMASE

The common *S. aureus* β-lactamases, which are responsible for penicillin resistance and are either plasmid- or transposon-encoded,[13-15] are substrate-inducible[16] and are regulated by a pathway homologous to the well-characterized β-lactamase induction pathway in *Bacillus*.[17] In early studies on *B. cereus* β-lactamase, it was observed that induction continued after free inducer was removed, suggesting that the inducer had become covalently bound to the bacteria.[18] Many years later, a transmembrane protein was identified that is required for β-lactamase induction in both *B. cereus* and *B. licheniformis*.[19] This protein, BlaR1, has a typical β-lactam binding pocket and does, indeed, bind penicillin covalently[20] so that the induction of β-lactamase in *Bacillus* parallels the covalent binding of β-lactams to the penicillin binding proteins involved in cell wall synthesis. In *S. aureus*, which expresses a plasmid-coded homologue of BlaR1,[21] induction does not continue following removal of the inducer (Ruzin A, Novick RP, unpublished data), suggesting that the inducer is not covalently bound to the BlaR1 homologue. In all cases, the net effect of activating BlaR1 is to relieve repression of the *blaZ* promoter by BlaI, which is assumed to bind the *blaZ* operator. The mechanism by which BlaR1 relieves repression is unclear; the BlaR1 proteins lack the standard histidine phosphokinase and nucleotide binding site motifs; however, they show significant similarity to the chemotaxis receptors,[22] which are methylated and act by modulating in turn

the level of methylation of an internal signal receptor, CheA. It is widely assumed that there is a third *bla* regulatory factor, BlaR2, on the basis of *bla* regulatory mutations in *B. cereus* and in *S. aureus* that are outside the standard *bla* region shown in Figure 9.1B. In *S. aureus*, mutations affecting this intermediate factor cause a high level constitutive phenotype,[17,23] whereas in *B. cereus*, they eliminate *bla* activity. This difference in phenotype may simply be a reflection of the different methods used to select the mutations rather than of any fundamental difference between the factors. There is currently considerable interest in finding this mythical gene, referred to as *blaR2*, not the least aspect of which is the possibility that *bla* induction involves a novel signal transduction mechanism. Expression of the methicillin resistance determinant, *mecA*, is regulated by a signal transduction mechanism homologous to that regulating *bla*; the signal transduction-*mecA* complex has evidently been formed by recombination between the *mecA* region and a co-resident *bla* regulon.[24] Additionally, *mecA* is repressed by an unlinked *bla* repressor and, remarkably, fully repressed strains are methicillin sensitive despite the fact that methicillin is an excellent inducer of the classical *bla* system.[16] Not surprisingly, many clinical MRSA isolates have escaped this regulation by an apparent insertion-deletion mutation eliminating *blaI* and part of *blaR1* and causing constitutive expression of *mecA*.[25]

Biologically, an interesting feature of *bla* resistance systems in staphylococci is that both the β-lactamase enzyme and the signal transduction receptor appear to have evolved from enzymes involved in cell wall biosynthesis and the *mecA* product is a novel member of the class of cell wall synthetic enzymes known as penicillin binding proteins (PBPs). Genetic considerations indicate that all of these resistance genes have been imported into the clinically relevant staphylococci from other unknown species by the agency of mobile genetic elements such as plasmids and transposons. Theoretically, they would have evolved de novo in these other species as part of the standard genetic complement.

4. LANTIBIOTICS

Lantibiotics are peptidic bacteriocins produced principally by Gram-positive bacteria, including lactobacilli (nisin) staphylococci (epidermin) and *B. subtilis* (subtilin). They are highly potent membranolytic toxins and are of general interest because they act on an extremely wide range of bacterial species. They are synthesized conventionally as precursors that are extensively modified posttranslationally. An N-terminal segment is removed, serine and threonine are dehydrated and the addition of sulfur from neighboring cysteine residues results in the formation of cyclic lanthionines. The various individual types are homologous and are synthesized and secreted by a complex multistep pathway consisting of two sets of genes, one containing the structural genes for synthesis and processing of the bacteriocins, the other consisting of two regulatory genes. In the case of nisin, and also subtilin, the regulatory genes have sequence motifs typical of the classical two-component signal transduction systems. In the case of epidermin, one of the genes in the regulatory operon, *epiQ*, resembles the signal transduction response regulators; the other resembles a serine protease rather than a histidine phosphokinase and so it is not clear whether epidermin synthesis is regulated by signal transduction. In the nisin pathway, the regulatory genes, *nisR*, and *nisK*, are directly downstream of the biosynthetic pathway[26] and probably belong to the same operon. Thus the nisin pathway is autocatalytic and it has recently been shown that nisin itself is the inducer (de Vos W, personal communication). In the epidermin pathway, the regulatory gene(s) are directly downstream of the biosynthetic operon but are transcribed in the opposite direction.[27] Their regulation is unknown.

Like other secondary metabolites, lantibiotics are typically produced postexponentially. If one assumes that their biological function is the destruction of other bacteria, and that the postexponential phase corresponds to the usual metabolic state of bacteria in the natural environment, then one might imagine lantibiotic production as a response to high population density. If so, then the inducing ligand might be produced in proportion to overall bacterial density.

5. THE *AGR* REGULON

S. aureus elaborates and secretes a large number of extracellular proteins, many of which are clearly involved in pathogenicity, others are suspected to be involved, and still others are known only as bands seen on SDS-PAGE analyses of culture supernatants. The best known staphylococcal exoproteins are listed in Table 9.2. These proteins can be placed in three regulatory classes: I, proteins that are produced and secreted during the postexponential phase in laboratory cultures; II, proteins that are produced during the exponential phase and are downregulated postexponentially; and III, proteins that are produced constitutively or in response to specific inducers, independently of the growth phase of the culture. The occurrence of pleiotropic exoprotein-deficient mutants,[28,29] plus the above-mentioned expression pattern, suggested coordinate regulation of exoprotein production, for classes I and II. However, as the Exp⁻ mutants showed a variety of phenotypes, with overlapping subsets of exoproteins showing loss of expression,[29] it was difficult to envision any simple genotypic mechanism; consequently, genetic analysis was initiated only when a transposon (Tn551) insertion causing an Exp⁻ phenotype became available.[1,30] This analysis resulted in the identification, cloning and sequencing of the *agr* locus, a global regulator of exoprotein production in *S. aureus*.[31,32]

The activity of *agr* mirrors the behavior of most of the proteins in classes I and II of Table 9.2. That is, most of the known proteins that are upregulated postexponentially are upregulated by *agr* and are not expressed by *agr*⁻ mutants, whereas those

Table 9.2. Regulatory classes of exoproteins in S. aureus

Class	Product[1]	
I. Produced post-exponentially and upregulated by *agr*	α-hemolysin	Serine protease
	β-hemolysin	Metallo protease
	δ-hemolysin	FAME
	γ-hemolysin	ETA
	leukocidin	ETB
	TSST-1	Staphylokinase
	SEB	
	SEC	
	SED	
II. Produced during exponential phase and downregulated by *agr*	Protein A	
	Coagulase	
	Fibronectin binding protein	
III. Produced constitutively and indifferent to *agr*	SEA	
	β-lactamase	
	Nuclease	

[1] TSST-1, toxic shock syndrome toxin 1; SEA, staphylococcal enterotoxin A; SEB, staphylococcal enterotoxin B; SEC, staphylococcal enterotoxin C; SED, staphylococcal enterotoxin D; FAME, fatty acid modifying enzyme; ETA, exfoliative toxin A; ETB, exfoliative toxin B.

that are downregulated postexponentially are downregulated by *agr* and are overexpressed by *agr⁻* mutants. And *agr⁻* rmutants are similar in phenotype to the commonest class of spontaneous Exp⁻ mutants.[28] In some cases, the latter can be complemented by some but not other *agr* subclones, suggesting that they are, actually, *agr* mutants. Others cannot be complemented by *agr*, indicating that other genes, outside of *agr*, are also required for exoprotein expression (see below).

5.1 THE *AGR* LOCUS

The *agr* locus has been analyzed independently by two groups[31-34] and the results of these analyses are in substantial agreement. As shown in Figure 9.1C, *agr* is a complex locus consisting of two divergent transcription units driven by promoters P2 and P3. The P2 transcript represents an operon containing four open reading frames, *agr A, B, C* and *D,* all four of which are required for *agr* function.[35] Two of these, *agrA* and *agrC,* correspond to the response regulator and histidine phosphokinase genes of the standard two component signal transduction system. AgrC

is predicted to be a transmembrane protein, from which it is inferred that the system is activated by an external ligand. Indeed, it has recently been observed that *S. aureus* cultures elaborate one or more soluble factors that activate transcription of the *agr* P3 operon (see below) immediately upon addition to a culture, i.e. before the system is normally activated.[36] Activity requires functional *agrA* and *agrC* genes, which is consistent with the possibility that these factors represent the putative ligand and that AgrC is the receptor. The activator is produced by strains lacking *agrA* and *agrC* but not by strains lacking *agrD* or *agrB,* suggesting that *agrD* and *B* either encode or regulate its production (Ji G, Novick RP, unpublished data). This observation brings *agr* activation into line with a number of other self-activating bacterial systems including the development of competence for transformation in *B. subtilis,*[37] conjugation in enterococci[38] and agrobacteria,[39] bioluminescence in *Vibrio* sp,[40] and elastase production in *Pseudomonas aeruginosa,*[41] all of which involve autoinducers that act via signal transduction pathways. A general principle that can be gleaned from these different systems is that all are mechanisms of sensing and responding to cell density. The biological need for sensing cell density, however, is not always clear. In the case of staphylococcal virulence, given that the hallmark of staphylococcal infection is the abscess and, assuming that the same mechanism operates in vivo, one might imagine a scheme such as that shown in Figure 9.2. In this model, the soluble virulence factors would not be needed until the tissue focus of infection becomes crowded and surrounded by antiinfective cells, fibrin and other localizing influences of the host. At this point, the bacteria will, in theory, be approaching stationary phase and will elaborate these factors in an attempt to breach the barriers by which the infective focus is contained. A role can readily be seen in such a scenario for leukocidin and the hemolytic (cytolytic) toxins, for proteases, nucleases, lipases, esterases, etc. The role of the superantigen toxins, in this scenario, seems less obvious, since these generally act at distant sites rather than locally. Their role in ablating V_β-specific sets of T-cells and inducing interleukins would clearly have an effect on the overall immunological and infective processes but would not seem to have any obvious local role.

In contrast, proteins that are downregulated by *agr*, such as protein A, coagulase, could be imagined to be more important early in infection, when the bacterial numbers are low and the incipient infection is more vulnerable to phagocytosis and other host defenses. Adhesins, such as fibronectin binding proteins, also downregulated by *agr*, can likewise be envisioned to be more important during the early stages of infection.

5.2 The Regulatory Circuit

Although the *agr* P2 operon seems to encode a signal transduction pathway, the actual intracellular effector of the *agr* response is specified by the P3 operon; the function of the P2 genes is to activate P2 and P3 transcription. Thus, activation of the P2 genes in the absence of the P3 region has no effect on exoprotein expression, whereas induction of P3 transcription in the absence of the P2 region initiates the entire *agr* response.[34,42,43] Remarkably, the P3 product that serves as the effector is the 514 nt P3 transcript itself, referred to as RNAIII, rather than any protein product.[42] RNAIII is a complex molecule that has several functions. Two major functions are transcriptional upregulation of 20 or more unlinked exoprotein genes and independent downregulation of at least three surface protein genes. The mechanism by which RNAIII regulates transcription is not known; however, it is likely that regulatory proteins, encoded outside of *agr* are involved. Several types of evidence are consistent with a role for regulatory proteins. First, there are no sequence elements in RNAIII or target genes that could support a direct RNA-DNA interaction involving the formation of a three-stranded structure by alternative base pairing rules[44-47] and, moreover, there is no precedent for any direct regulatory RNA-DNA interaction. Second, we have isolated mutations in genes outside of *agr* that are required for the function of RNAIII. That is, strains containing such mutations produce wildtype RNAIII but do not show the *agr* response (Novick RP, unpublished data). Third, we have observed that treatment with protein synthesis inhibitors, such as erythromycin (Em) or chloramphenicol (Cm) mimics the effects of RNAIII. That is, transcription of genes that are upregulated by *agr*, such as *hla*, is strongly stimulated by Em or Cm whereas transcription of genes that are downregulated by *agr*, such as *spa*, is sharply inhibited by Em or Cm. These effects, which are seen in an *agr*-null as well as in an *agr*⁺ background (submitted for publication), suggest the involvement of labile regulatory proteins that decay in the presence of Em or Cm. For *hla*, the putative regulatory protein would be a repressor, for *spa*, an activator. Either the function or synthesis of these intermediary regulatory

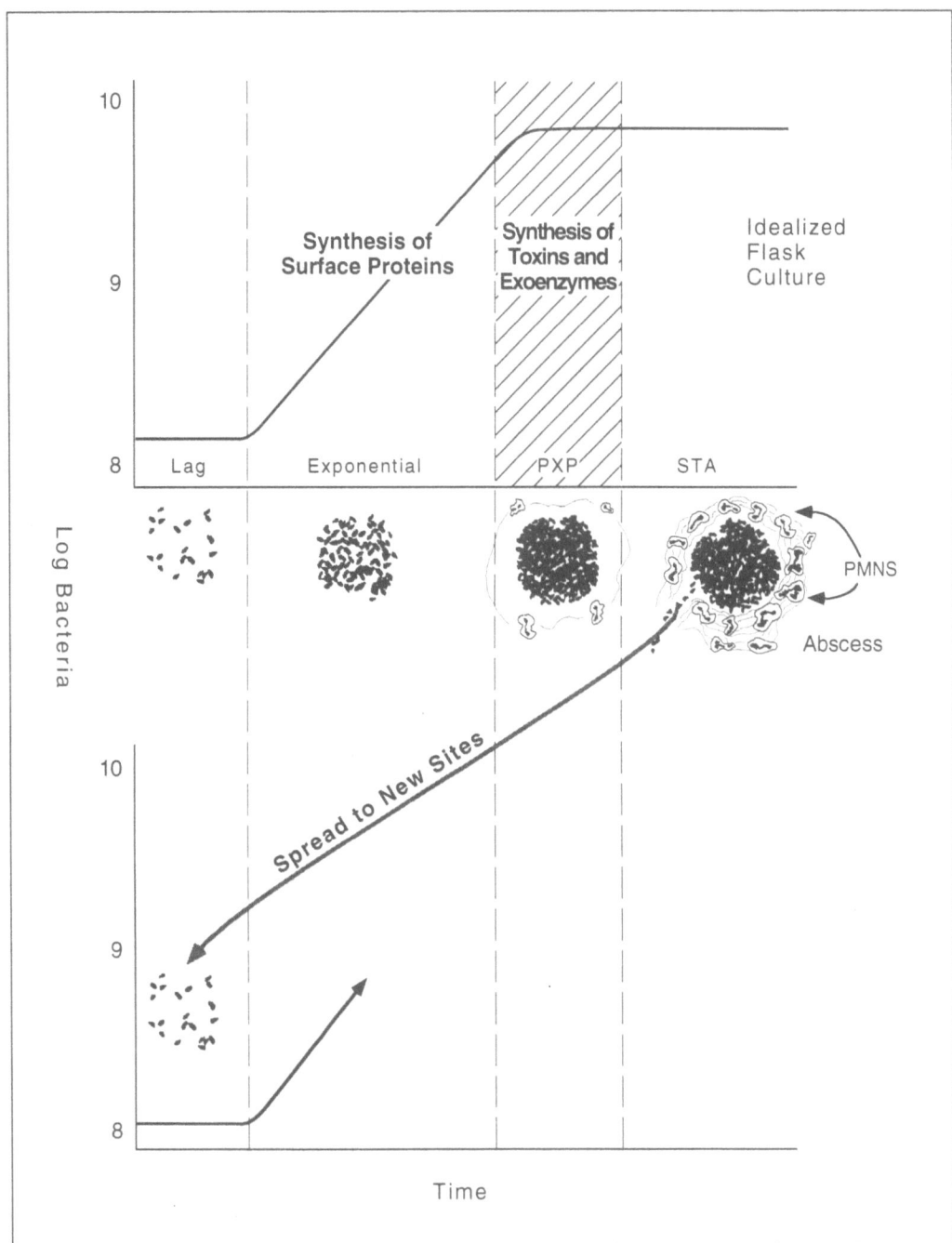

Fig. 9.2. Hypothetical role of virulence factors in staphylococcal infection. In this model, bacteria in lag phase would initiate the infection and would then enter exponential phase where multiplication would begin, accompanied by synthesis of surface proteins. Crowding during postexponential phase would activate a density sensing mechanism which would upregulate toxic exoproteins, enabling the organisms to escape from the incipient abscess and spread to new sites, where the cycle would be repeated.

proteins would then be the target of RNAIII. It has been observed that novobiocin at subinhibitory concentrations (that have been observed in other systems to increase negative supercoiling) induces the transcription of at least one *agr*-regulated gene, *eta*.[48] This suggests that increased superhelix density may be involved and the possibility must be considered that the putative labile regulators may be histone-like proteins analogous to H-NS.

Recently, two additional genes, unlinked to *agr*, have been identified by transposon (Tn917) insertions causing a pleiotropic Exp⁻ phenotype. One of these, *sarA*,[49,50] homologous to *Shigella flexneri virF* may be a second transcriptional activator required for the activation of the *agr* P2 promoter. Interestingly, whereas *agr* defective mutants show a considerable but not total reduction in mouse virulence,[51] an *agr-sar* double mutant is totally avirulent.[52] Evidently the overproduction of protein A partially compensates for the loss of other virulence factors in *agr⁻* mutants, whereas the *agr-sar* double mutant does not produce detectable protein A. The site of action of the other gene, *xpr*[53] is unknown. The existence of these genes indicates that exoprotein regulation in *S. aureus* possesses additional levels of complexity and that much additional work remains to be done.

RNAIII also affects the translation of at least some of the *agr*-regulated products, independently of its effects on transcription. The best studied example is α-hemolysin. Deletion of the first 92 nt of RNAIII eliminates translation of α-hemolysin but has no effect on the regulation of transcription by RNAIII. Within the deleted region are sequences showing strong complementarity to the 5' end of the *hla* mRNA untranslated leader: the obvious possibility is that in the absence of RNAIII, the *hla* leader folds into an untranslatable configuration and that pairing between these two sets of complementary sequences is required for *hla* mRNA translation (see Fig. 9.1C). Which, if any, other exoprotein genes are similarly regulated is not presently known.

5.3 TARGET SEQUENCES

A conserved 5' sequence has been identified in a number of exoprotein genes (Projan S, personal communication). Preliminary genetic analysis has suggested that this element may be involved in *agr*-specific regulation of *eta* transcription.[48] Deletions caused a reduction in *eta* expression and in some but not all cases affected the *agr*-specific activity of the promoter. The overall picture that emerges is that the in vitro regulation of virulence factor expression in *S. aureus* is a multistage process in which one can discern several stages that correspond to the biological scheme shown in Figure 9.2. In early exponential phase surface proteins are expressed strongly, exoproteins are not expressed; *agr* transcription is at a low basal rate. During mid-exponential phase, *agr* activator accumulates to a threshold level, *agr* transcription is activated, which results in auto-amplification of the *agr* transcripts and products, and in the shut-off of surface gene transcription. In postexponential phase, a second signal is generated which is required jointly with *agr* RNAIII for the transcriptional upregulation of exoprotein genes. The combination of these two signals causes exoprotein genes to be upregulated and their products to be produced in large quantities. In stationary phase, *agr* transcription stops, exoprotein synthesis stops, and stationary phase promoters are presumably activated.

Elaboration of virulence factors in a multi-stage sequence has also been envisioned for *Bordetella pertussis* and is described in chapter 2.

6. STREPTOCOCCI

In enterococci, the pheromone-mediated conjugation mechanism has a number of features that imply the existence of signal transduction pathways. Small peptides (pheromones) produced by the potential recipient organism induce the potential donor to elaborate a surface protein, aggregation substance (AS) that causes agglutination. It is within the resulting clumps, which include recipient as well as donor organisms, that DNA transfer takes place. There is suggestive evidence that this clumping factor may be an adhesin[54] suggesting that the mating pathway may have a role in virulence. The signal transduction mechanism here seems quite different from that of the standard two-component systems. A surface protein, TraC, evidently binds the pheromone and presents it to an oligopeptide transport system by which it is internalized and interacts directly with *traA*, the gene encoding AS. This system may thus resemble the pathways for activation of sporulation and competence in *B. subtilis*, in which small regulatory peptides are internalized by a specific uptake systems.[37] In other words, signal transduction may sometimes involve the transport of the signal itself rather than of an activating impulse initiated by it.

In group A streptococci, a regulatory gene *mry*[55] or *virR*[56] controls the synthesis of the M protein, a surface protein important for pathogenicity, and of several other virulence-associated proteins, including the Ig-Fc binding proteins and the antichemotactic complement-inactivating C5a peptidase, as well.[57] The *mry/virR* product resembles the response-regulators of the classical two-component systems, except that it has duplicate motifs. No gene corresponding to the histidine kinase component has been identified. Nevertheless, environmental factors such as CO_2[58] and anaerobiosis[59] affect expression of these genes.

7. CONCLUDING COMMENTS

Though the extent of signal transduction in GPC is only just beginning to be appreciated, its role in virulence is likely to be as important, or possibly even more important than in Gram-negative bacteria. This is because Gram-positive bacteria cause disease primarily by means of extracellular virulence factors; as it is a general rule that synthesis of extracellular proteins is regulated by signal transduction, it stands to reason that virulence regulons will be so regulated, from which it may be inferred that substances in in-

fected tissue may serve as key inducing ligands. Future studies on virulence regulons in these species will reveal the validity of this concept.

REFERENCES

1. Recsei P, Kreswirth B, O'Reilly M et al. Regulation of exoprotein gene expression by *agr*. Mol Gen Genet 1986; 202:58-61.
2. Novick RP, Richmond MH. Nature and interactions of the genetic elements governing penicillinase synthesis in *Staphylococcus aureus*. J Bacteriol 1965; 90:467-480.
3. Wang P-Z, Projan SJ, Novick RP. Nucleotide sequence of β-lactamase regulatory genes from staphylococcal plasmid pI258. Nucl Acids Res 1991; 19:4000.
4. Arthur M, Molinas C, Courvalin P. The VanS-VanR two-component regulatory system controls synthesis of depsipeptide peptidoglycan precursors in *Enterococcus faecium*. J Bacteriol 1992; 174:2582-2591.
5. Bayles KW. The use of degenerate, sensor gene-specific, oligodeoxyribonucleotide primers to amplify DNA fragments from *Staphylococcus aureus*. Gene 1993; 123:99-103.
6. Handwerger S, Pucci MJ, Volk KJ et al. The cytoplasmic peptidoglycan precursor of vancomycin-resistant *Enterococcus faecalis* terminates in lactate. J Bacteriol 1992; 174:5982-4.
7. Wright GD, Molinas C, Arthur M et al. Characterization of VanY, a DD-carboxypeptidase from vancomycin resistant *Enterococcus faecium* BM4147. Antimicrob Agents Chemother 1992; 36:1514-1518.
8. Liu J, Volk KJ, Lee MS, Pucci MJ, Handwerger S. Binding studies of vancomycin to the cytoplasmic peptidoglycan precursors by affinity capillary electrophoresis. (Submitted).
9. Stock JB, Ninfa AJ, Stock AM. Protein phosphorylation and regulation of adaptive responses in bacteria. Microbiol Rev 1989; 53:450-490.
10. Handwerger S, Pucci MJ, Kolokathis A. Vancomycin resistance is encoded on a

pheromone responsive plasmid in *Enterococcus faecium* 228. Antimicrob Agents Chemother 1990; 34:358-60.

11. Arthur M, Molinas C, Depardieu F et al. Characterization of Tn1546, a Tn3-related transposon conferring glycopeptide resistance by synthesis of depsipeptide peptidoglycan precursors in *Enterococcus faecium* BM4147. J Bacteriol 1993; 175:117-27.

12. Handwerger S, Pucci MJ, Volk KJ et al. Vancomycin-resistant *Leuconostoc mesenteroides* and *Lactobacillus casei* synthesize cytoplasmic peptidoglycan precursors that terminate in lactate. J Bacteriol 1994; 176:260-264.

13. Novick RP. Analysis by transduction of mutations affecting penicillinase formation in *Staphylococcus aureus*. J Gen Microbiol 1963; 33:121-136.

14. Asheshov EH. The genetics of penicillinase production in *Staphylococcus aureus* strain PS80. J Gen Microbiol 1969; 59:289-301.

15. Rowland SJ, Dyke KG. Tn552, a novel transposable element from *Staphylococcus aureus*. Mol Microbiol 1990; 4:961-975.

16. Novick RP. Staphylococcal penicillinase and the new penicillins. Biochem J 1962; 83:229-235.

17. Collins JF. The *Bacillus lichenformis* β-lactamase system. Beta-lactamases, eds. J.M.T. Hamilton Miller, and J.T. Smith. (London: Academic Press, 1979) 351-368.

18. Pollock MR, Perret CJ. The relation between fixation of penicillin sulphur and penicillinase adaptation in *B. cereus*. The British J Exp Pathol 1951; 32:387-396.

19. Kobayashi T, Zhu JE, Nicholls NJ et al. A second regulatory gene, *blaR1*, encoding a potential penicillin-binding protein required for induction of β-lactamase in *Bacillus licheniformis*. J Bacteriol 1987; 169:3873-3878.

20. Zhu Y, Englebert S, Joris B et al. Structure, function, and fate of the BlaR signal transducer involved in induction of β-lactamase in *Bacillus licheniformis*. J Bacteriol 1992; 174:6171-6178.

21. Wang P-Z, Projan SJ, Novick RP. Nucleotide sequence of β-lactamase regulatory genes from staphylococcal plasmid pI258. Nucl Acids Res 1991; 19:4000.

22. Joris B, Ledent P, Kobayashi T et al. Expression in *Eschericia coli* of the 346-carboxy terminal domain of the BlaR sensory transducer protein of *Bacillus licheniformis* as a water-soluble 26,000 Mr penicillin-binding protein. FEMS Lett 1990; 70:107-114.

23. Cohen S, Vernon EG, Sweeney HM. Differential depression of staphylococcal plasmid and chromosomal penicillinase genes by a class of unlinked chromosomal mutations (R2⁻). J Bacteriol 1970; 103:616-621.

24. Matsuhashi M et al. Molecular and genetic studies on methicillin resistance in staphylococci. In: Novick RP, ed. Molecular biology of the staphylococci. New York: VCH Publishers, Inc, 1990: 457-470.

25. Archer GL et al. Dissemination among staphylococci of DNA sequences associated with methicillin resistance. Antimicrobial Agents and Chemotherapy 1994; 38:447-454.

26. Engelke G, Gutowski-Eckel Z, Kiesau P et al. Regulation of nisin biosynthesis and immunity in *Lactococcus lactis* 6F3. Appl Environ Microbiol 1994; 60:814-25.

27. Peschel A, Augustin J, Kupke T et al. Regulation of epidermin biosynthetic genes by EpiQ. Mol Microbiol 1993; 9:31-39.

28. Bjorklind A, Arvidson S. Mutants of *Staphylococcus aureus* affected in the regulation of exoprotein synthesis. FEMS Microbiol Lett 1980; 7:203-206.

29. McClatchy JK, Rosenblum ED. Biological properties of alpha toxin mutants of *Staphylococcus aureus*. J Bacteriol 1966; 92:575-579.

30. Mallonee DH, Glatz BA, Pattee P. Chromosomal mapping of a gene affecting enterotoxin A production in *Staphylococcus aureus*. Appl Environ Microbiol 1982; 43:397-402.

31. Peng H-L, Novick RP, Kreiswirth B et al. Cloning, characterization and sequencing of an accessory gene regulator (*agr*) in *Staphylococcus aureus*. J Bacteriol 1988; 170:4365-4372.

32. Morfeldt E, Jazon L, Arvidson S et al. Cloning of a chromosomal locus (*exp*) which regulates the expression of several exoprotein genes in *Staphylococcus aureus*. Mol Gen Genet 1988; 211:435-440.

33. Janzon L, Lofdahl S, Arvidson S. Identification and nucleotide sequence of the delta-lysin gene, *hld*, adjacent to the accessory gene regulator (*agr*) of *Staphylococcus aureus*. Mol Gen Genet 1989; 219:480-485.

34. Kornblum J et al. Agr: a polycistronic locus regulating exoprotein synthesis in *Staphylococcus aureus*. In: Novick RP, ed. Molecular Biology of the Staphylococci. New York: VCH Publishers, 1990:373-402.

35. Novick RP, et al. The *agr* P2 operon: an autocatalytic sensory transduction system in *Staphylococcus aureus*. Mol Gen Genet 1995; (In press)

36. Balaban N, Novick RP. Autocrine regulation of exoprotein synthesis in *Staphylococcus aureus*. Proc Natl Acad Sci USA.

37. Magnuson R, Solomon J, Grossman AD. Biochemical and genetic characterization of a competence pheromone from *B. subtilis*. Cell 1994; 77:207-216.

38. Clewell DB. Bacterial sex pheromone-induced plasmid transfer. Cell 1993; 73:9-12.

39. Fuqua WC, Winans SC. A LuxR-LuxI type regulatory system activates Agrobectarium Ti plasmid conjugal transfer in the presence of a plant tumor metabolite. J Bacteriol 1994; 176:2796-806.

40. Meighen EA, Dunlap PV. Physiological, biochemical and genetic control of bacterial bioluminescence. Advances in Microbial Physiology 1993; 34: 1-67.

41. Passador L, Cook JM, Gambello MJ et al. Expression of *Pseudomonas aeroginosa* virulence genes requires cell-to-cell communication. Science 1993; 250:1127-1130.

42. Novick RP, Ross HF, Projan SJ et al. Synthesis of staphylococcal virulence factors is controlled by a regulatory RNA molecule. EMBO J 1993; 12:3967-3975.

43. Janzon L, Arvidson S. The role of the delta-lysin gene (*hld*) in the regulation of virulence genes by the accessory gene regulator (*agr*) in *Staphylococcus aureus*. EMBO J; 1990; 9:1391-1399.

44. Collier DA, Wells RD. Effect of length, supercoiling, and pH on intramolecular triplex formation. J Biol Chem 1990; 265:10652-10658.

45. Griffin LC, Dervan PB. Recognition of thymine-adenine base pairs by guanine in a pyrimidine triple helix motif. Science 1989; 245:967-971.

46. Maher LJ III, Dervan PB, Wold BJ. Kinetic analysis of oligodeoxyribonucleotide-directed triple-helix formation on DNA. Biochemistry 1990; 29:8820-8826.

47. Wells RD et al. Unusual DNA structures and the probes used for their detection. In: Wells RD, Harvey SC, eds. Unusual DNA Structures. New York: Springer-Verlag, 1987:1-21.

48. Sheehan BJ. The regulation of epidermolytic toxin A expression in *Staphylococcus aureus*. Doctoral dissertation. University of Dublin, 1992.

49. Cheung AL et al. Regulation of exoprotein expression in *Staphylococcus aureus* by a locus (*sar*) distinct from *agr*. Proc Natl Acad Sci USA 1992; 89:6462-6466.

50. Cheung AL, Projan SJ. Cloning and sequencing of *sarA* of *Staphylococcus aureus*, a gene required for the expression of *agr*. J Bacteriol 1994; 176:4168-72.

51. Foster TJ et al. Genetic studies of virulence factors of *Staphylococcus aureus*. Properties of coagulase and gamma-toxin, alpha-toxin, beta-toxin and protein A in the pathogenesis of *S. aureus* infections. In: Novick RP, ed. Molecular Biology of the Staphylococci, New York: VCH Publishers, 1990:403-420.

52. Cheung AL et al. Diminished virulence of *sar⁻/agr⁻* mutant of *Staphylococcus aureus* in the rabbit model of endocarditis. The Journal of Clinical Investigation 1994; 94:000-000.

53 Hart ME, Smeltzer MS, Iandolo JJ. The extracellular protein regulator (*xpr*) affects exoprotein and *agr* mRNA levels in *Staphylococcus aureus*. J Bacteriol 1993; 175:7875-7879.

54. Chow JW, Thal LA, Perri MB et al. Plasmid-associated hemolysin and aggregation

substance production contribute to virulence in experimental enterococcal endocarditis. Antimicrob Agents Chemother 1993; 37:2474-2477.

55. Perez-Casal J, Caparon MG, Scott JR. Mry, a trans-acting positive regulator of the M protein gene of *Streptococcus pyogenes* with similarity to the receptor proteins of two-component regulatory systems. J Bacteriol 1991; 173:2617-24.

56. Chen C, Bormann N, Cleary PP. VirR and Mry are homologous trans-acting regulators of M protein and C5a peptidase expression in group A streptococci. Mol Gen Genet 1993; 241:685-693.

57. Podbielski A. Three different types of organization of the *vir* regulon in group A streptococci, Mol Gen Genet 1993; 237:287-300.

58. Okada N, Geist RT, Caparon MG. Positive transcriptional control of *mry* regulates virulence in the group A streptococcus. Mol Microbiol 1993; 7:893-903.

59. Podbielski A, Peterson JA, Cleary P. Surface protein-CAT reporter fusions demonstrate differential gene expression in the *vir* regulon of *Streptococcus pyogenes*. Mol Microbiol 1992; 6:2253-2265.

INDEX

Page numbers in italics denote figures (f) or tables (t).

MOLECULAR BIOLOGY
INTELLIGENCE UNIT
AVAILABLE AND UPCOMING TITLES

Neuroscience Intelligence Unit

Available and Upcoming Titles

☐ Neurodegenerative Diseases and Mitochondrial
Metabolism
M. Flint Beal, Harvard University

☐ Molecular and Cellular Mechanisms of Neostriatum
Marjorie A. Ariano and D. James Surmeier,
Chicago Medical School

☐ Ca^{2+} Regulation in Neurodegenerative Disorders
Claus W. Heizmann and Katharin Braun,
Kinderspital-Zürich

☐ Measuring Movement and Locomotion:
From Invertebrates to Humans
Klaus-Peter Ossenkopp, Martin Kavaliers and
Paul Sanberg, University of Western Ontario and
University of South Florida

☐ Triple Repeats in Inherited Neurologic Disease
Henry Epstein, University of Texas-Houston

☐ Cholecystokinin and Anxiety
Jacques Bradwejn, McGill University

☐ Neurofilament Structure and Function
Gerry Shaw, University of Florida

☐ Molecular and Functional Biology
of Neurotropic Factors
Karoly Nikolics, Genentech

☐ Prion-related Encephalopathies:
Molecular Mechanisms
Gianluigi Forloni, Istituto di Ricerche Farmacologiche
"Mario Negri"-Milan

☐ Neurotoxins and Ion Channels
Alan Harvey, A.J. Anderson and E.G. Rowan,
University of Strathclyde

☐ Analysis and Modeling of the Mammalian Cortex
Malcolm P. Young, University of Oxford

☐ Free Radical Metabolism and Brain Dysfunction
Irène Ceballos-Picot, Hôpital Necker-Paris

☐ Molecular Mechanisms of the Action
of Benzodiazepines
Adam Doble and Ian L. Martin, Rhône-Poulenc Rorer
and University of Alberta

☐ Neurodevelopmental Hypothesis of Schizophrenia
John L. Waddington and Peter Buckley,
Royal College of Surgeons-Ireland

☐ Synaptic Plasticity in the Retina
H.J. Wagner, Mustafa Djamgoz and Reto Weiler,
University of Tübingen

☐ Non-classical Properties of Acetylcholine
Margaret Appleyard, Royal Free Hospital-London

☐ Molecular Mechanisms of Segmental Patterning
in the Vertebrate Nervous System
David G. Wilkinson, National Institute
of Medical Research, United Kingdom

☐ Molecular Character of Memory
in the Prefrontal Cortex
Fraser Wilson, Yale University

MEDICAL INTELLIGENCE UNIT

AVAILABLE AND UPCOMING TITLES

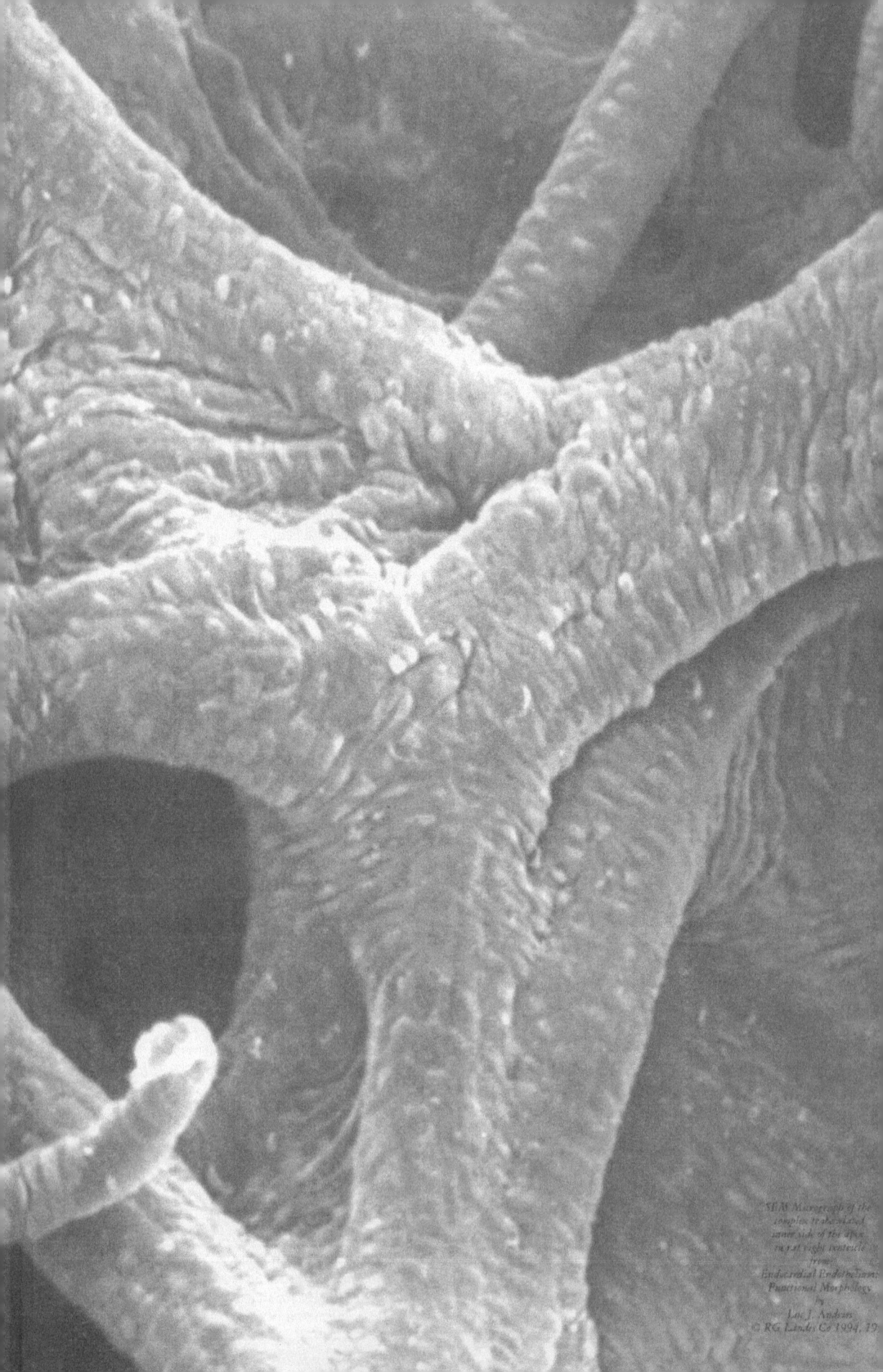

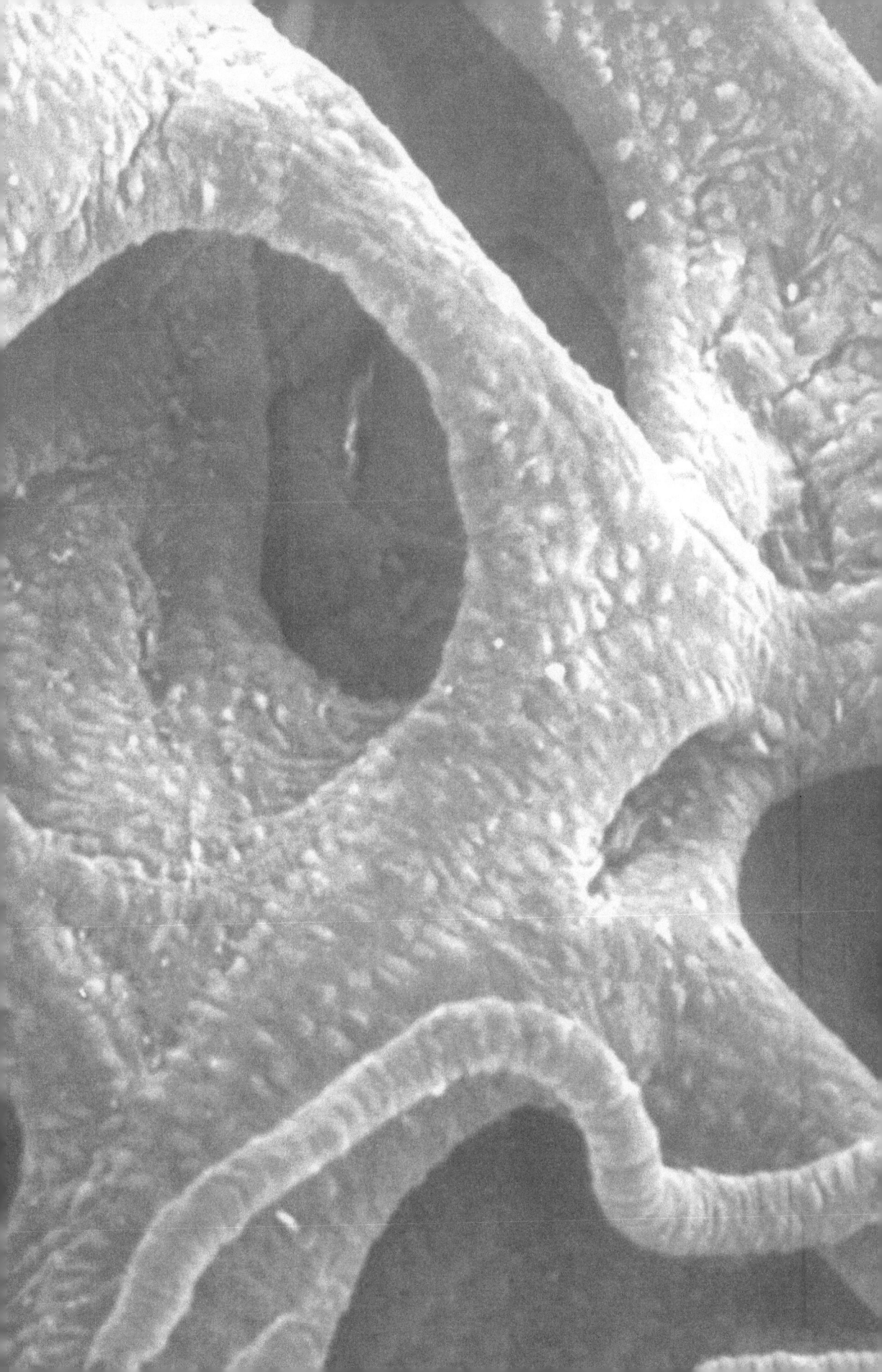